THE GREAT DIGITIZATION

AND THE QUEST TO KNOW EVERYTHING

LUCIEN X. POLASTRON

TRANSLATED BY JON E. GRAHAM

Inner Traditions
Rochester, Vermont

Inner Traditions
One Park Street
Rochester, Vermont 05767
www.InnerTraditions.com

Originally published in French by Éditions Denoël under the title *La Grande Numérisation*
First U.S. edition published in 2009 by Inner Traditions

Library of Congress Cataloging-in-Publication Data

Polastron, Lucien X.
[Grande numérisation. English]
The great digitization and the quest to know everything / Lucien X. Polastron ; translated by Jon E. Graham. — 1st U.S. ed.
p. cm.
Includes bibliographical references and index.
ISBN 978-1-59477-243-6 (pbk.)
1. Books and reading. 2. Books—Digitization. 3. Libraries—Forecasting. 4. Communication in learning and scholarship—Technological innovations. I. Title.
Z1003.P747513 2009
028.7—dc22

2008048425

Printed and bound in the United States by Lake Book Manufacturing

10 9 8 7 6 5 4 3 2 1

Text design and layout by Priscilla Baker
This book was typeset in Garamond Premier Pro with Avenir and Bank Gothic used as display typefaces

Inner Traditions wishes to express its appreciation for assistance given by the government of France through the National Book Office of the Ministère de la Culture in the preparation of this translation.

Nous tenons à exprimer nos plus vifs remerciements au government de la France et le ministère de la Culture, Centre National du Livre, pour leur concours dans le préparation de la traduction de cet ouvrage.

Contents

Translator's Preface

With the pace of change in the world of computers and communication more likely to be measured in nanoseconds than years, many readers may wonder why we would choose to translate and publish a book that is destined to be out of date, in some respects, even before the edition hits the stores.

The answer is really quite simple: in this book, Lucien X. Polastron is raising questions that no one else has yet raised about the future of how we obtain knowledge—both in our use of books and public libraries as well as the great digital revolution that is all around us. Casting his historian's-eye view on a still-unfolding scenario gives him a distance to observe the world's embrace of this new technology that threatens to irrevocably alter not only the way we handle information but also transform the way our very thought is structured—the very way we think. As the subtitle of his French edition asks: "Is there thought after paper?"

The revolution offered by digitization, Polastron points out, should not be compared with the revolution created by the Gutenberg printing press, which simply provided a mechanical means to quickly and efficiently reproduce manuscripts that until that time had to be laboriously created by hand. Rather, he says, this digital revolution is on a par with that of the replacement of the scroll by the codex, which occurred around AD 300 and which resulted in entirely new reading habits. The scroll offered a continuous reading experience, whereas the codex required a text to be dismembered into discreet pages, thus incorporating a tic in the process when the reader is forced to turn the page and reengage with the words he is reading. Reading on the computer likewise is developing entirely new habits for the eye and the mind behind it.

Polastron explores the pros and cons the digital revolution offers. He first asks: Will the individual become acclimated to a lazier form of pointillist thinking, developed from reading digests and snippets that can at best graze consciousness and which reduce the ability for the kind of deep contemplation that is required and fostered by the slow digestion of a book? When one reads a book, one is holding the totality of the author's thoughts, arguments, and thesis in hand—its beginning, middle, and end—and there is an expectation when one begins a book that there will, in fact, be a beginning, middle, and end. Or, Polastron ponders, will the digitizing of books result in opportunities for a more holistic understanding of a given subject, made possible by the transformation of the work into an endless text with multiple branches easily accessible through the ability to click on a given word in one text and thereby gain access to another text, and another—every word of a text becoming the potential platform of another reality? The jury is still out, obviously, but a change of such magnitude should be given the kind of serious study that a mad rush to the future eschews.

Another possible result of digitization Polastron explores has both moral and political implications. It is an issue that puts librarians on the front line, just as they were in protesting the invasive powers the U.S. authorities granted themselves with the Patriot Act to investigate any individual's reading habits without their knowing. Polastron points out that the very notion of free public reading may be an unanticipated casualty of digitization as many institutions—both public and private—are construing the addition of a layer of software to a text, even those in the public domain, as giving them ownership and control over that text. This should be deeply disturbing to any citizen of a democracy, given that this form of government can only flourish when its citizens are well educated and well informed. Polastron asks: Will the digitization of books open up mass accessibility to the literature of the world or create a closed, elitist, and totalitarian system of information retrieval? Will it affect knowledge in the future? Quoting librarians around the world, he identifies a critical point—that "information is not knowledge."

Although Polastron touches on all the enormous technical and moral issues involved in digitization, he begins his investigation with a discussion of the immense funds spent on the new home for the National Library of France (BnF). He does so to demonstrate that there is a dynamic tension between the physicality of library buildings and their physical book collections and the movement to digitize books. And while the discussion uses France's national library as the example, the problems and situation facing that library has its counterparts in all the great libraries of the English-speaking world, including the Library of Congress in the United States and the British Library in England.

Digitization is an epiphenomenon of globalization—its impact will be felt throughout the world, although the nature of that impact may take various shapes in response to regional and national character traits. The problems faced by the National Library of France are already in evidence in cities across the United States and Britain. For example, the San Francisco Library, which cost $126 million—money that was no longer available to buy new books—created an even more distressing situation when it was discovered that the new facility lacked the proper space for the three million books held in its former home, the "Old Main." This led to the library's administrators discreetly destroying the overflow—estimated to be anywhere from two hundred thousand to five hundred thousand books. Far from serving as a cautionary tale to other libraries, the British Library was next to be caught in the act of ordering the destruction of part of its collection when its facilities were renovated and "upgraded." While vast sums are being spent on new quarters for these institutions, the funding for their ambitious plans to digitize their collections remains far short of what is required for this mammoth project. The irony of new library facilities resulting in the destruction of physical books, presumably prior to their being digitized, is a mockery of the goal to make these collections more accessible to the public.

Polastron is raising the red flags that no one else is talking about. *The Great Digitizaton* presents a lucid and fascinating look at a terrain that is in transition. It explores the quest for a universal library, the function of the book, the role of librarians, and how the future of information is going to be structured and controlled. Polastron warns us that with the future of digitization, knowledge itself might be at risk because it will no longer be free.

Jon E. Graham
January 2009

Jon E. Graham has translated more than thirty books from French, including the award-winning *Books on Fire* by Lucien Polastron, *The Secret Message of Jules Verne* by Michel Lamy, and *Immaculate Conception* by André Breton and Paul Eluard. Mr. Graham's translation of Jean-Luc Steinmetz's *Arthur Rimbaud: Presence of an Enigma* was selected by the *Los Angeles Times* as one of the ten best translations of 2000. As a graphic artist, his works have been displayed in galleries in New York, Philadelphia, San Diego, Seattle, Vancouver, Paris, Prague, Stockholm, and the National Art Museum of Portugal in Lisbon. His illustrations have appeared in various books and magazines. He is the acquisitions editor for Inner Traditions International, a position he has held since 1996.

Acknowledgments

The author would like to express his gratitude here to all those who contributed their time, knowledge, and experience to this investigation, particularly Alain Absire, Corinne Aribaud, Michèle Battisti, Luc Bellier, Gérard Cathaly-Prétoi, Antoine Chéron, Pierre Cubaud, Boris Engelson, Michel Fingerhut, Alain Jacqueson, Catherine Lupovici, Pierre Mansiat, Guillaume Marsal, John L. Needham, Ève Netchine, Alexis Rivier, Jean-Marc Rod, Laurence Santantonios, Michel Valensi, and Eric White, as well as many others—you know who you are.

INTRODUCTION

Finding Information or Funding a National Library?

All expressions from all times and all places are mixed together in the same, constantly reworked hypertext, from where they are sent on at any time, everywhere.

MANUEL CASTELLS, *FIN DE MILLÉNARE* FROM *L'ÈRE DE L'INFORMATION*

We waver between the illusion of completion and the vertigo of the ungraspable.

In the name of completion, we like to think that there is a unique order enabling us from the outset to attain knowledge; in the name of the ungraspable, we like to think that order and disorder are two words that mean chance.

GEORGES PEREC, "NOTES BRÈVES SUR L'ART ET MANIÈRE DE RANGER SES LIVRES" IN *PENSER/CLASSER*

The future of gathering knowledge is now a few years old. It's time to think about what this revolution really means and therefore come back to how it all started in 2005: the sheer shock of civilizations that occurred between the New World and the Old, between Mountain View and Paris in France: a conflict of totally different views on what a global library is made of and for. Then we suddenly had to face the following question: Do thoughts belong to pages, pages to books, and books to a nation? It was the war of pure ideas against concrete.

Let us speak in terms of francs, because that gives us a round figure we will not obtain using euros.

The upkeep of the academic library nowadays known as the National Library of France costs one billion francs a year—in other words, one-fifth of the subsidies given public cultural establishments, seven times more than it received at the beginning in the 1990s, although the number of its users has barely tripled.* It is also evident that the institution has acquired fewer and fewer books during this same span of time. During this same period the British Library bought 140,000 titles, as opposed to the 60,000 purchased by its counterpart in France. So where is all the money going? A large portion of it goes to expenses connected with the building of Tolbiac, which underwrites several deluxe troupes, a fire brigade, a complete police station, the cleaners of the giant windowpanes of the library complex, forest wardens, and other antitheft squealers who have been added to the payroll. You must add to this all the other professional bodies essential for the survival and development of your standard immense national library. Graduate students are already occupying 58 percent of the places allotted for professionals on the basement floor (which have been deserted by the writers) and 84 percent of those that had been Mitterandically conceived for the "underprivileged" occupy the ground floor. These statistics are good grounds for asking if so much capital would not be better spent divvied up among the nation's uni-

*Page 54 of the 2003 BnF activity report explicitly estimates an average of 1,175 daily visitors to the former premises on rue Richelieu, versus 3,441 to the new establishment of the Tolbiac area.

versity libraries (the sum of 150 million euros represents three-fifths of what they receive). This tidy sum could be dedicated to the digitization in text mode of books that are useful to the researchers who work more and more on screen and whose attendance in the dungeons of the Thirteenth Arrondissement continues to shrink. However, "the sophisticated nature of the technical and data-processing equipment there requires careful upkeep and the cost can only grow higher by virtue of its rapid obsolescence and limited life span," as was noted in a Senate report published in 2000.[1]

Jacques Attali, an individual who is regarded as a leprechaun because of his habit of tossing off ideas that could pass as daring because they rarely fall back to earth, has been proved right, at least on one point.* In 1988, he valiantly strove to bring about the digitization of the National Library's entire collection. By all evidence, if the immense fortune swallowed up by the construction of Tolbiac alone (1.2 billion euros) had been devoted to this daunting task, the effort would now be past history. It is likely that if the library had remained headquartered on the rue Richelieu, its reading rooms would have again become suitable for flesh-and-blood researchers. Its premises were regarded as a poetic spot quite in keeping with our era's fetishistic and nostalgic trends. Members of the international scholarly community would have been delighted to occasionally reward themselves with an instructive sojourn spent in this temple of paper worshippers.

Fifteen years later, it has been discovered with much blushing among the librarians reduced to chair attendants that too little has been done—too little, too poorly, and too late.

The polemic was accidentally relaunched in January 2005, just as it was sinking into oblivion: Paris had hopped aboard the train of change

*An observation with which he seems to be in complete agreement, based on his letter published in the March 14, 2005, issue of *L'Express.* [Jacques Attali is an economist and author, and was an adviser to former French president François Mitterand. —*Trans.*]

fully expecting it would reach its destination. However, it has not traveled very far:

> In 1995 at the BNF, for example, we had endeavored to grasp (in partnership with the School of Engineering and the most promising research engine of the time, Altavista's LiveTopics, developed by the Frenchman François Bourdoncle) just what was actually involved in "finding information" within a structured collection of different forms of knowledge, like the Gallica collection. It is too bad that this initiative was never followed up. The Exalead engine created by this same François Bourdoncle is undoubtedly still a candidate for renewed efforts ten years later.

So recalled Yannick Maignien, who led the scientific digitization policy of the National Library of France until 1997.

There is no question that Gallica, the online library inaugurated in January 2000, offers the opportunity to read scanned old documents at no charge:

> This collection of novels, essays, magazines, famous texts, and rarer works gives every reader the means, whether he is merely curious or a bibliophile, whether she is a high school or college student, to acquire a more extensive understanding of an era in its political, philosophical, scientific, or literary aspects. (BnF document online)

Launched in 1992, the digitization program was expected to grow by 100,000 titles every three years, only half of which could be consulted without any formalities in one's home (the rest, still under copyright, could be read only in the reading room). We are still painfully shy of this target. Only 52,764 monographs and 13,482 volumes of periodicals are currently available, as the process proved to be both slower and more expensive than anticipated: 9 million euros were spent from

1992 to 1998, then 500,000 to 750,000 euros a year, according to inside sources—a fairly pathetic figure. "The management of the institution wondered if an opportunity therefore existed to charge user fees for certain services," Le Hir wrote in 2002,* but perhaps it was works under copyright it had in mind. The case file is unclear, when you read elsewhere that the texts online represent only 55 percent of the work that has been digitized but not yet put online. A quick calculation tells you that 130,000 books have a virtual double today. Information appears neither in the establishment's 2003 activity report—in which Gallica is barely mentioned—nor in the "documentary charter" published in autumn 2005[2] for the purpose of making things clearer.

Now, with the exception of the 1,200 works in "text mode" provided by the CNRS—which already has 3,737 on its own Frantext and whose use, even for paying customers, is forbidden to those dubious citizens lacking an institutional stamp of approval†—none of these extravagant reproductions of admirable editions is either very useful to the specialist or very easy to unearth. These reproductions are simply photographs of pages that are revealed at the speed of an old turtle to the eyes of the stubborn individual. That is, if he or she already knows the site of the archive. And without any index or possible citation, one must copy by hand or print out the book in A4 format, page after page, through the OCR mill, which will stumble over every slightly blurred serif of the Bodoni and meticulously include every speck of fly crap. This cheap anachronism is called "image mode." Realists

*Pierre Le Hir, in "The BnF Is Digitizing Its Rare Collections for Easier Use" (*Le Monde,* March 7, 2002), cites the figure of 750,000 euros, which subsequently fell to 500,000. But suddenly, in 2005, the institution complained of having at its disposal only "a thousand times less" than Google, which would be no more than 123,421 euros. Everyone knows that literary people are easily thrown by numbers. However, if this figure is confirmed, it is not a European library we need to organize but a charity sale.

†Could "CNRS" [Centre National de la Recherche Scientifique, National Center for Scientific Research —*Trans.*] be an incomplete anagram for *caserne* [barracks]? Fortunately, less rigid entities do exist. Since 1994, Eulogos in Rome has performed marvelous work using the same criteria, which can be freely discovered with Infratext (all the addresses for digitalized collections are listed in the back of the book).

coldly maintain today: "In an information society, text can only exist as text mode.* It is better to have less text, but text that is still real text, than more text frozen into image mode."[3] Everything takes place as if Gallica had been defined by besieged soldiers, although the more likely designers were a bibliophile consortium that was worthy of esteem, but whose piercing vision encompassed only the distant past. France therefore displays its written holdings slowly and majestically, starting with texts that have often been made outdated by new critical publications. In a word, all the progress made here is merely putting under glass what had been formerly sitting beneath a bell jar.

Transformed into a sanctuary as exuberant as the Hôtel des Invalides,† Gallica is an admirable souvenir album that the well-educated father can leaf through electronically before perhaps ordering that strange and wild text from his dumbfounded bookseller. Precise studies of its users[4]—on average, forty-eight-year-old urban males—have clearly shown that they can be primarily divided into professionals "looking for basic information" and "bibliophiles anticipating a purchase." Those polled who admitted having read the text amounted to one of every sixteen, and even this was less a reading than a "consultation/navigation."

Although bands of impatient volunteers willing to do the work‡ are

*Special software, still being perfected, will be capable of transforming the page of a book in the Gallica collection into a page of text on demand. For more on this topic, read Elisabeth Freyre, "Structuration et description automatique des documents électroniques: le projet européen METAe" (*Culture et Recherche* no. 100, first trimester, 2004). In this case, why doesn't the archive apply itself to the entire corpus without dilly-dallying?

†[Site of Napoleon's tomb. —*Trans.*]

‡For example, http://ebooksgratuits.com/index.php has been offering (since quite recently) complete access to Machiavelli's very amusing novella, *Belfagor, the Devil Who Took a Wife,* in two translations, 1664 and 1825, with a sober note from the cyber benefactor who took on the task of transforming it into four formats—Mobipocket, PDF, Word, and HTML—so that its contents would be available to everyone on any reading engine. It should be noted, incidentally, that the philosophy of the group responsible for this collection of approximately a thousand free titles is that "the work of digitization, OCR scanning, and correction performed on these texts does not bestow any form of copyright upon them." Could a fair and just perception of the new technology contribute to the restoration of the sense of honor?

beginning to sprout up everywhere, it goes without saying that the first healthy decision for the BnF to take would be to complete the task it has undertaken and duplicate the texts of its collection with real texts, then cast them into the void to be sniffed out by everybody at the far ends of the earth. There would be no need for anyone to know the password beforehand. Equipped with an electronic tattoo, these digitized books could only increase the institution's notoriety tenfold.

Thereupon an earthquake: Google announces its contracts to digitize the books of five major American and English libraries. The dead emerged from their tombs and the French national archive from its august autism, although it needed some six weeks to hone its opinion. Six years might have been better, but this may not have made it any the wiser.

1

BnF versus BNF*

As early as 1994, foreign observers were guffawing at the sight of what they had always known and respected as the "BN" [Bibliothéque nationale] becoming, in the maelstrom of all the incompetence surrounding its modernization, the "BNF" [Bibliothéque nationale de France (National Library of France)]—in other words, a title given a double-layered coating of red, white, and blue paint. One can imagine the chuckles if the Library of Congress were to suddenly add a similar construction to its name and become the "Federal Library of the Congress of the United States." How could we forget that the second-largest library in the Western world soberly describes itself as "British"? The Parisian shift, though, owes nothing to driving blind: the name of the digitization program, Gallica, was already a rusting cock-a-doodle-do.† So the response to Google—which was not expecting one—was quite predictable: culture found itself wrapped in the flag.

---

*We write it out "BnF" due to a stylistic affectation in the logo. In the real world, typographical religion demands that the three letters be uppercase (with periods, if you please, adds an emendator). But as it seems to me that there is already enough *national* in this matter, let's adopt the logo the young institution wants while clinging to the hope that future reforms will bring back the sober, elegantly pompous, and definitive Library of France. One must always be more royalist than the king.

†[The rooster is a symbol of France. —*Trans.*]

No more than any other high official can the spokesperson of an organ as emblematic as the BnF* accept the responsibility of justifying earlier mistakes, profound as those mistakes might be. Born in a family of highly placed state servants since at least the time of Clemenceau,† the most estimable man who assumed this position of responsibility for the institution in 2005 could hardly imagine any better argument for sovereignty to brandish than one as ludicrous and indefensible as this:

> I remember the experience of the Bicentennial of the Revolution in 1989 when I was in charge of its celebrations. It would have been detrimental and detestable to the nation's equilibrium, and to the image and knowledge it has of itself, of its past, and of the events—both dark and light—that it was our duty to commemorate, to have sought only in English or American databases for a narrative and interpretation, a narrative deformed by so many biases: *The Scarlet Pimpernel* crushing *Quatre-vingt-treize,* valiant British aristocrats triumphing over bloody Jacobins, the guillotine obscuring the rights of man and the dazzling institutions of the Convention.[1]

It would have been child's play to come up with a retort against this then—but what newspaper, radio, or political adversary would have had the knowledge and the cheek to do so?—pointing out that on the one hand it was a distorted, narrow, and paternalistic vision of the past; and on the other, that American universities are overflowing with impeccable and incredible collections of completely digitized documents from which a student can draw forth the very reality of French history, such as the *quasi totality* of the bills and proceedings of the

*Jean-Noël Jeanneney was its "president," but the French researcher has great difficulty explaining to his Canadian or Chinese colleagues why this country has such a wealth of presidents outside the Élysée and that, no, it is not the sole source of most of our problems, nor especially of our inability to resolve them.

†[Georges Clemenceau (1841–1929), French statesman who was prime minister of France several times, including during the First World War, and played a major role in the drafting of the Treaty of Versailles. —*Trans.*]

French Revolutionary government.[2] As for "America's crushing domination" over language, even the modestly informed are aware that less than half the Library of Congress collection is in English, just as in the collections making up the Google Five (as the partner libraries in this project found themselves nicknamed), only 49 percent of the material is in English, as opposed to 23 percent in French, German, and Spanish, with the remainder divvied up among four hundred other idioms.

Although Montesquieu was born in La Brède and Perrault in Paris, it is thanks to Richelieu or to the hooligan Napoleon that a wealth of books has piled up in this territory. Their nationalization in no way authorizes any nationalist exploitation—quite the contrary. When revolutionaries are striving for the common good, it is the human race they have in mind, and it has not escaped any apprentice scholar that true knowledge knows no frontier. (I'd like to point out, inasmuch as it seems necessary to remind people, that even some of our most classic authors churned out hymns of this nature on occasion.) If knowledge is synonymous with freedom, it is because it guarantees access to a vision of life without restrictions: "The term *National Literature* does not mean much today; we are heading toward an era of *Universal Literature,* and everyone should strive to hasten the advent of this era," as Goethe, having just finished reading a Chinese novel, remarked at dinner on Wednesday, January 31, 1827.[3] Haven't we taken even one step further in the 180 intervening years, or do we truly want to go back two centuries?

Added to this irksome concoction is an unpleasant pinch of judicial legalese. The public today has been well enough drilled about copyright matters to know that the contents of books whose authors died before 1930 no longer belong to anyone. That is to say that the best of humanity's achievements throughout the whole of its history and possibly sufficient on their own for the salvation of civilization have no ownership. Consequently, nothing authorizes the employees, paid by the taxpayers, to be the keepers of their physical appearance—*and to facilitate their reading*—to speak in their name. Any attempt at rhetorical seizure of these texts is infinitely less tolerable than that of rapacious capitalists on the Web. Compared to their simple and frank logic, such an attempt resembles a ciborium theft by the beadle. But here, as stated on the Gallica site, is a shortcoming that is even worse:

> The National Library of France is the holder of author copyrights on the Gallica site. The reproduction of this site's content for strictly personal use is free. In the context of communications, publishing, or other activities of a professional nature, only short citations are authorized on the condition that mention is given of BnF/Gallica. Any other reproduction or representation, whole or substantial, of this site's content by any kind of procedure requires express permission of the BnF.

How is it that bureaucrats—book specialists—who appeal so readily to the notion of nation, and who should therefore be paragons of intellectual generosity and ethics times three—dare talk of "author copyrights" with respect to a technological task inherent to the ordinary discharge of the duties imparted to them? I will revisit this painful question a little later, but a public service would certainly open itself to criticism if it allowed itself to be dragged into the state of the new (and most unappetizing) mind-set that consists of creating a private copyright out of public property by the simple application of a layer of software. Less repugnant—so to speak—is the image of the soldier of fortune planting the tattered flag of his native land on the soil of an isle that until then had been paradise.

Research engine, Web, "digitization," blog aggregator, syndication, and hundreds of other concepts were and still are forbidding abstractions for the majority of people, caught short by an invasion of technicians and merchants who are, as we shall see, the Ostrogoth hordes of today. Is it necessary to take advantage of this anxiety to sow panic and condemn the specter under whose feet the French language will never sprout again? Consulted, the most minor literary figure even slightly brushed by the algorithm could have objected that the worst that could happen in the beginning would be that a lazy user looking for just any book to browse through would eventually stumble upon Philip Roth or even Faulkner before, let's say, André Maurois. This affair would not have kicked up such a fuss if someone hadn't struck the key in the keyhole of the collective unconscious and cracked open the unwholesome cupboard filled with inferiority complexes, illiterate French-speaking communities, anti-Hollywood paranoia, and the defense of genuine Camembert.

The bugle's echo roused the Élysée Palace, whose occupant then was still snared in some dreadful political predicament and saw this as "manna from heaven."[4] Despite his ignorance of the subject, the president got his oversize hackles up, telling himself that Bill Gates would support him in the event of an error, formed ranks with a government delighted to have a spot of diversion, and ordered the march forward to more chimeras such as the cultural Euro-pudding. Tolbiac began playing the grasshopper and began knocking on the doors of the neighboring ants seeking to get their digitalized resources put at common disposition. The poorest of these neighbors, Estonia, Lithuania, Slovenia, and Slovakia, signed first. The British phoned in their encouragement.

The very reactions that should have been avoided in the planetary poker game dealt and played perfectly by Google were exactly what occurred. An initial loss of nerve was followed by the stupefied rigidity that was caused by stumbling over the principle of a rooted culture that no longer exists except in the crossed eyes of veteran extreme right-wingers with the inability to see that an opportunity was arising in a positively magical manner. A little tact would have enabled the BnF to make wide use of this opportunity with little or no expense to itself, while providing a positive influence. This influence could have countered some of its more unfriendly aspects and allowed it to establish itself as an honorary engine.

Instead, due to the puerile ambience of a world war—the split among the nations of the world by George Bush's "you are either for us or against us" invasion of Iraq—we caved and are still caving under American jibes. We notice the knowing smiles in London and Barcelona when the name of Louis de Funès* is mentioned, and the total perplexity in Japan. Character assassinations flew through the Internet, more civilly expressed by the librarians† and more violently by the anonymous critics, to the great consternation of the average curators and librarians when they pulled their noses out of the *Bulletin des bibliothèques de France* [Bulletin of the Libraries of France]. On the other hand, the French press gave a very warm welcome to this brand-new subject of digitization, about which no one understood a thing, but which offered a triple winner: supposed modernity, defense of the culture, and a Gaullist nostalgia so typical of the 1960s. One lone journalist had the bad taste to specifically point out a close resemblance to certain grandiose and characteristic disasters that this era had in plenty: Bertrand Le Gendre, "Le plan calcul de la BnF" [The Plan Calcul of the BnF], *Le Monde,* April 22, 2005. His article was never cited in the interminable and repetitious press review later published by this institution.

Among the corpses the future will count when studying this battlefield, it will be noted that one of the most commonly shared conclusions by foreign blogs a year later was the very negative "Places like France have been extremely critical of Google Print."‡ "Extremely," in this instance, clearly means too much.

Although locked up in impassive police custody, books will continue to have their own thoughts on the matter. For example, they will still think, even with regard to high officials, that a bureaucrat is a temporary organism. How can a brain that is constantly torn over the

*[Louis de Funès (1914–1983) was a French actor and comedian, well known for his slapstick performances as hyperactive and bad-tempered middle-aged men. —*Trans.*]

†For example, Karen G. Schneider, *Free Range Librarian,* February 20, 2005—http://freerangelibrarian.com/archives/022005/toutalorsthefren.php—or Mark Liberman, *Language Log,* February 20, 2005—http://itre.cis.upenn.edu/myl/languagelog/archives/001911.html.

‡[In English in the original. —*Trans.*]

renewal of its mandate, pipe breakages, and union harassment grasp the true measure of eternity?

From the depths of the library rises the lament of the martyr:

*Shaken in my seat by the steps of readers rattling the ever-crooked floorboards, I was gazing at the poor crest of trees lit by the failing light of day. I contemplate this library given life by 2,099 salaried employees and placed between four dangers: the Seine, financial scandal, patriotism, and author copyright. One day, devoured by the demons pressing upon it, land speculation will repossess it like a property that had been overrun against its will.**

*Cf. François-René Chateaubriand, *Les Martyrs.*

2

For Wells Is Not the Plural of Orwell

At age seventy, the Martian ambassador, greatly alarmed by the state of planet Earth and with an eye to helping humanity escape its own "growing powers of waste and destruction," became convinced that nothing could be more effective than a general and permanent distribution of universal knowledge: encyclopedism. In the footprints of Comenius, "Everyone should learn everything," and Diderot, "If we want philosophers to move forward, we need to bring the people to where the philosophers are now," Wells suddenly put all his time and energy into working on the most extravagant idea.[1] This is precisely what we see happening today beneath the fires of current events.

The formidable collection of all forms of knowledge—an immeasurably vaster project than the *Encyclopaedia Britannica* would ever have dreamed of tackling, one employing thousands perpetually with an eye to daily and international coordination of everything that might emerge from intellectual institutions—was a concept H. G. Wells (1866–1946) first formulated in a chapter of *The Work, Wealth and Happiness of Mankind* in 1931. He then applied it in three novels—*The Croquet Player, The Camford Visitation,* and *Star-Begotten*—and in a work entitled *World Brain,* which synthesized a vision that has become quite familiar to us recently.

World Brain is an anthology of five conferences, four articles, and a radio address, dating from 1936 to 1938. The first was entitled "World Encyclopedia." It was delivered on November 20, 1936, at the Royal Institution of Great Britain and advocated for a worldwide neoencyclopedic movement that would display its supremacy over all governments and other bodies hostile to the development of individual consciousness. As we are reminded by one of his biographers, Wells had already foreseen for more than ten years "that the inconsistencies of politicians, the concessions made to the electoral body, the trend of following public opinion instead of guiding it, will cast an increasingly serious discredit upon Western democracies." He also originated the idea of an "open conspiracy" against "spiritual confusion, the lack of courage, indolence, and a wasteful selfishness."[2]

Contrary to that, at a time when a journalist like Diderot could bring together select famous contributors to compile articles, "for gentlemen by gentlemen," that were truly devoid of propaganda and advertising, Wells said the new encyclopedia should be elaborated by highly qualified authorities and updated *at the same pace* as the advance of research in every domain. This should be the "mental background of every intelligent man," an "organization that would extend like a nervous system supervising the world, a network connecting all intellectual workers through their shared interests and by a common means of expression, tending to become an increasingly conscious cooperative unit."

A year later, in 1937, the notion was defined more explicitly in the Brain Organization of the Modern World conference, held in the United States:

> Our world is changing, and it is changing with an ever increasing violence. An old world dies about us. A new world struggles into existence. But it is not developing the brain and the sensitiveness and

> delicacy necessary for its new life. . . . Two hundred years ago Oliver Goldsmith said that if every time a man fired a gun in England, someone was killed in China and we should never hear of it, and no one would bother very much about it. All that is changed. We should hear about that murdered Chinaman at once.

This "imperative expansion of the scale of the community in which we have to live" should be used to our benefit by placing the whole of our ideas and knowledge in a network; "this encyclopedic organization need not now be concentrated in one place" nor be materialized in a row of volumes that have been printed and distributed once and for all. What other means besides printing could be used, then? As the electronic offshoot known as the Internet had not yet come into the most imaginative mind, it was microfilm that carried the perspectives of Utopia in 1937. All that was needed was the support to be provided by cheap, mass-produced, standardized projectors so that any student, in any part of the world, would be able to sit with his projector in his study at his or her convenience to examine *any* book and *any* document, in an exact replica.

With the totalitarian shadow already hovering during these years, such an arrangement would furthermore allow the exact and integral duplication of the world brain "in Peru, China, Iceland, Central Africa, or wherever else seems to afford an insurance against danger and interruption. It can have at once, the concentration of a craniate animal and the diffused vitality of an amoeba."

If 1938 Europe had not suddenly accelerated its race to catastrophe, Wells could well have witnessed the fulfillment of his plan. The congress of August 19 and 20, 1937, that he organized in Paris under the patronage of an embryo of Unesco, brought together precursors (Paul Otlet, inventor of the Mundaneum; G. W. Davies, founder of the American Documentation Institute; and so forth) and donors, mainly American, such as the Kodak Company, whose mouth was already watering. Whatever it may have been, we can see that the dream imposed a serious weight from the onset: the Wellsian brain would have involved a centralized structure that would have been administratively overburdened and fraught with dangers. This is how *The Shape of Things to Come* (1933) describes a permanent world encyclopedia headquartered in Barcelona: it "collects, classifies, preserves in order, and makes available all that is known" while employing seventeen million people in every country of the world.

Instead of which, we have the libertarian system of the Web—lurking just beneath the surface of Wells's desire—that pulses like an amoeba in everyone's home and nowhere simultaneously; and while it is easy to make money from it, censors and bureaucrats struggle to control it. It is up to everyone to find his or her own way, which is a good thing. But this avatar of the humanist dream bearing peace and harmony thanks to the manna from science could well lead, in opposition to Wells's *World Brain,* to the emergence of a *Global Brain,*[3] cobbled together on a daily basis by the user. We would then be talking about, rather, what some people consider to be a communal brain pudding.

In the future "there will not be an illiterate left in the world. There will hardly be a single person who is uninformed or misinformed." Herbert George Wells apparently sincerely believed in the potential benefits of the collective effort of intelligent individuals toward increasing intelligence. George Orwell found him naive.

3

Geneva: World Capital

Another librarial Utopia that was run aground by European self-destructiveness preceded Wells's endeavor* by ten years. This ideal was more classical in conception, offering the originality of being associated with the objectives of Le Corbusier, whose mind and relatively elaborate plans were not devoid of poetry, nor failed to insidiously evoke some recent construction.† This detail alone makes it worth one's while to blow away the thick layer of dust slumbering on this file.

The son of a rich Belgian financier nicknamed "the emperor of the tramways," lawyer Paul Otlet (1868–1944) nurtured a passion for bibliography, invented the notion of bibliology, and, starting in 1895, dedicated himself not only to his idea of pacifism through the encyclopedia, but

---

*This is one of the most hard-headed of all human desires, one on which enthusiasts of varied personalities have exhausted their energies. Before Paul Otlet, there was Eugene Godin (1856–1942), who devoted his life to the public reading room of the future national library in Paris. A relentless proponent of public education, he created *L'Encyclopédie nationale* as "a shared information office that should permit each user to inform his neighbor on the subjects of his expertise. It was founded on a solid principle: 'Everything is in everyone.' The 100,000 index cards Godin collected were bequeathed by his widow to the institution that today no longer knows what became of them." Ève Netchine, "Un 'hydropath' au service de la lecture publique à la Bibliothèque nationale," *Histoire des bibliothécaires* [History of Librarians], Presses de l'ENSSIB, Lyon: 2006.

†[Otlet commissioned Le Corbusier to design the Geneva Mundaneum, which was never built. —*Trans.*]

also to the elaboration of the UDC, the Universal Decimal Classification, which improves upon the Dewey system's classification of the different branches of learning—which is to say, books—when it proves necessary to arrange them in some kind of order.

His library? It would be a global one: "Just as universal thought is one, all the books in which it finds its expansion are all the elements of a large, ideal universal Book. And the private libraries housing these books are all part of a Library that can be considered universal in spirit as well, although it is divided into multiple collections." Fed by the legal deposit (books that have to be sent to the Library of Congress when taking out copyright or its equivalent in other nations) of every country, this entity would not be satisfied by mere books but was intended to hold all diplomas, laws, archives, disks, films, and photographs there as well. The World Palace would house this collection. Alas, the scorn, and then hostility, of the authorities quickly transformed this institution into the "wandering Jew of documentation." Manhandled during the rough moving of its contents, it was finally shut down in 1934 as a "den of Bolsheviks." To the right-wing press, Otlet was never anything but the "general penpusher," whereas the Brussels weekly *Le Rouge et le Noir* designated him "the clerk who sought not to betray."

This is the reason that "it was proposed to establish the Mundaneum in Geneva, with all forces as one. This institution was intended to be Headquarters for International Associations, Congress, and Free Movements, as well as a Scientific Center serving both educational and documentary needs, one that would realize on a global scale and with the cooperation of official bodies the five great traditional institutions of Intellectual Labor: Library, Museum, Scientific Associations, University, and Institute." Studded as it is with commas and capital letters, Paul Otlet's act of faith constitutes number 128 of the publication of the Union of International Associations, World Palace, Brussels, August 1928. Poignantly, the cover of this pamphlet hopes "to commemorate in 1930 ten years of efforts toward peace and collaboration."[1]

Perhaps Le Corbusier did not read Otlet's text to the end, because his building would not have housed even a hundredth of the collection the latter envisioned. He imagined a ground floor on pilings "with spiraling automobile circulation beneath" and, most important, "a building for work and not for pomp and ceremony." Le Corbusier is an architect.*

> The interior of the vast prism topping the pilings is entirely empty from top to bottom and between its walls. The space will be systematically occupied by metal installations of footbridges, shelving, toboggans, elevators, and pneumatics.
>
> The elevators and visitors' ramp are enclosed in a glass funnel. This way the visitor ascending to the reading rooms will be able to take stock during his journey of just what an international library and its organization consists.
>
> The reading rooms, both large and small, are at the top of the building. The administrative offices will also be located there along with the coatroom, the restaurant, and the roof garden, with its shelters and walkways overlooking the entire site.
>
> The Library . . . stands smooth and rectilinear, with its mute walls topped by a sparkling crown of windows in the middle of a grass lawn.

The architect's stated intention was to erect the Mundaneum between Grand-Saconnex and Pregny to provide "the most majestic sights in the four cardinal directions." Concluded Otlet, taking advantage of the opportunity to use two additional capital letters, "Thus on one spot of the Globe, the image and entire meaning of the World can be perceived and understood."

*The master builder reserved his lyricism for the museum, a marriage of the Aztec pyramid with the ziggurat girded by a 2.5-kilometer spiral slope that allowed the top of the building to be reached from the outside. This shape was not by chance. The collections of the World Museum follow a chronological order from the sky to the ground, starting from prehistory (at the top there is nothing but a skull on view) and growing larger as one descends. "What a philosophy is revealed to whoever understands this!"

But the exact opposite occurred. Instead of a "cultural boundary," there was a monument to bureaucracy. Destiny erected over this same inspired hill the League of Nations, the Palace of Nations, and the United Nations when *nation* is the first word that needs to be erased to ensure the emergence of a Utopia. Every human being knows—and sometimes even suffers from concretely—the unique ineffectiveness of this authority as it presents itself today; there is no need to dwell on the subject here. But the new institution was unable to agree on one of the 377 architectural plans submitted for its inaugural design. Eventually the authors of five of these plans—one Italian, a Swiss, two French, and a Hungarian, all unknown—were chosen to work together. They were able to get along well enough to come up with a building that could have earned the admiration of a Mussolini or of a small fry like Ceauşescu.

Who listens to the visionaries? Paul Otlet announced: "One can also imagine the electric telescope that will permit one to read in the comfort of one's home books that are exposed in the 'teleg' room of the large libraries, turned to pages that have been requested in advance. This will be the telephoned book." And he added in the Treatise of Documentation: Telelecture shall "provide texts for long-distance reading" and telescription shall "add inscriptions to existing texts from a distance." Enough said.

4

Quick, Quick

When H. G. Wells elaborated the total library that would be perpetually updated, the responsibility for feeding it was something he awarded only to the most competent specialists.

A reality based in popular science fiction is today attempting to fulfill this desire: in 1994, the first pedestrians on the Internet immediately imagined an adaptable virtual encyclopedia to whose wealth everyone would add (but keep in mind that "everyone" then meant scientists). The concept was baptized, without much ado, Interpedia. They imagined that because it was free of any financial constraint, one day this innovation could not help but be superior to what existed among the publishers of paper books. Being virtual, it had no limitations that could prevent it from replacing all forms of knowledge. Ten years later, this bubble had turned into a snowball. It calls itself Wikipedia.

"The name Wiki comes from an adjective in the Hawaiian language: *wiki wiki,* which means fast. . . . The principle is simple: it is a cooperative model for document writing. Practically speaking, any visitor has an opportunity to change the page he is reading. . . . For example, a first author can write an article, a second completes it, then a visitor corrects any potential errors he may have noted" (source, Wikipedia). Founded by Jimmy Wales and Larry Sanger, Wikipedia was born on January 15, 2001, in English. Five years later, the inflatable encyclopedia existed in more than two hundred languages, of which close to half contained one hundred articles. There are close to a million in English and two hundred thousand in French,* the language that was fourth to appear, following German and Japanese. These figures are constantly growing, but the sphere's accounting is well maintained.

Robert McHenry, former editor in chief of the Encyclopaedia Britannica, had the means to size up the new venture, as well as the itch to critique it.[1] He had every opportunity to note that in the first definition on the English site, the adjective *reliable* came well after "free, and largest in depth and breadth." It is very clear and straightforward: the encyclopedia is constructed more by amateurs than professionals; if mistakes abound, they will be corrected over time through the increase of the number of visitors to the site. This perpetual improvement is therefore "quasi-Darwinian." Over the long haul, the true statements would outlive the erroneous, although new details would be flooding in each day. "All that is required is to believe in it," grumbles McHenry, who titles his article "The Faith-Based Encyclopedia." But he was perhaps wrong to appeal to reason, and it is a waste of time to point this out without correcting the thousands of false statements and dates. To have a clear conscience, the author of these lines rectified one article on a famous subject and two known to be urban legends, if not refuted outright by historians. A year later, not one of my changes had been touched.

The truth about Wikipedia may be that the giant is not intended so much to be used as to use. Seeing things for the best, which should be understood à la Borges, the entity does not welcome any reader, only players, people who amuse themselves playing encyclopedia just as little boys once played with Erector sets at the home of their wealthiest neighbor. Furthermore, the participants have rapidly established, with papal-like gravity, a hierarchical structure of stewards, bureaucrats,

*[As of mid-August 2007, there are now close to two million articles in the English version of Wikipedia (note that on the FAQ page it says 1.4 million) and more than half a million on the French site. —*Trans.*]

developers, and administrators; social relationships are codified with plus and minus signs, and double-plus and -minus signs.

These players boil over with passion. When one French contributor to Wikipedia got indigestion from the *Britannica* criticism, he immediately developed the argument on his personal site that Diderot's Encyclopedia received contributions from such rabble-rousers as Voltaire, Rousseau, and Montesquieu. An undertaking like this one could only be political, he said, and even revolutionary. Consequently, according to him, *Britannica* has no right to the label of encyclopedia; only Wikipedia is thus authorized, because this resource is enriched by "a free man helping to build a free society."[2] This same Internaut, in response to the observations appearing in newspapers about the system's unreliability, laid out fervent arguments studded with grammatical errors in French. Don't get hung up on this last detail; according to the *wiki* philosophy, every reader will in turn correct the errors he uncovers—if he sees them. True enough, but when deficient documentation is displayed like a barrage of bullets against reasoning . . .

Anyone who wishes to make up his own mind about this intriguing question should read the fastidious debates that follow the displays for and against. The enthusiast of subjective logorrhea will find what it takes to become intoxicated here, along with the opportunity to spill his own contribution into the stream. Similarly, the impartial observer will observe that the most frequently employed argument here is also the most biased—to wit, that traditional encyclopedias never corrected mistakes. But it would undoubtedly do no good to point out that the number and scope of these latter are minuscule in comparison to the disjointed praise, and that their rectification is real, albeit quite slow—but this is not what is expected of you.

The arguments on both sides of the debate are equal in weight—both the detractors and the supporters of this online encyclopedia concept have valid points. One might feel something disturbing, perhaps a whiff from the confessional, in this verbose hyperexcitation at the disposal of an apparently unstoppable logic. Fists clenched as they pound their keyboards and with eyes glued to their monitors, are these avowed partisans of collective encyclopedism going to start playing the *Revenge of the Nibelungen*?

More simply, the founder, influenced at the end of 2005 by criticisms targeting inadmissible errors bordering on racist disinformation found on the site, seemed to direct himself toward dividing his successful venture into two separate entities: on one side, a large planetary rough draft open to all contributions, on the other his definitive version authenticated by indisputable leading authorities, which is a return to Wells.*

All utopias have an end, but at least this one experienced the beginning of a realization.

*[Additional fuel for this criticism was provided by the ABC program *Good Morning America* in August 2007, when reporter John Berman broke the story that corporations like Wal-Mart and Starbucks were editing the articles devoted to them on Wikipedia. —*Trans.*]

5

Volutes

"From Marrakech to Mandalay, the world smokes Du Maurier," declares the ad for an elegant Art Deco cigarette in an old *Vogue,* the substratum, to some extent, of an archaeology of globalization. In this slogan, one could replace "smoke" with "devour" for a multitude of countries where the library is underweight, or has been pillaged or burned—as happened to the Heinrich Heine collection at the Goethe Institute in Lome at the end of April 2005 ("There where one burns books* . . .")—and has only a meager chance of one day filling its shelves: the dazzling slowness of the Alexandrina can serve as a signpost in this regard. In addition to *Jamaica Inn* in several languages—if the reader selects Daphne, as in Du Maurier, the last name of eleven authors—millions of books will soon open their pages in the most remote regions of the world,† provided that the material

*The complete phrase and its analysis can be found in my *Books on Fire;* it recurs with such stubborn persistence in the daily news that a new book was necessary: *Feux sans fin* [Endless Fires], scheduled to be published sometime in the near future by Denoël.

†Free machines are also spreading rapidly. One such example is the computer with a starting crank, of which MIT plans to produce several hundred million as part of the campaign "One Laptop per Child" for those countries least apt to equip themselves on this massive scale. Stripped of all superfluous elements, these machines operate on Linux, not Windows, and can connect to the Internet by wireless. They cost no more than one hundred dollars to manufacture (which incidentally sheds an interesting light on our own three-thousand-dollar machines). Source: http://laptop.media.mit.edu.

and the connection can make the rendezvous. There is no real reason to worry about the hardware—it becomes less expensive and performs better by the day—and for the rest, it depends on the regime: whether it is a banana republic, fascist, or enlightened. On a whole, and so long as the opportunity remains free, this is incontestably book-manna that should rain down on the poor. No adversary of the Great Digitization should dare deny that this represents a peerless, fast, and unhoped-for remedy to the deplorable state of the libraries in those lands where we maintain chaos.

And what does one foresee for the more privileged? Many researchers are rejoicing at the prospect of spending less time on public transportation, and calm will be restored to the reading rooms. But what else? Only those who have a personal stake in the matter refuse, despite their constantly declining sales figures, to acknowledge that the general system of hard-copy book production reached the end of its roll around the year 2000, and that it will no longer be possible to continue adding useless books every month when there are already more useful ones than we need: the Library of Congress surpassed more than six hundred miles of full shelves around 1990. In France, libraries are finally admitting in hushed tones that they are ridding themselves of books. "With discernment," states Claudine Belayche,[1] but by the armload: 105,221 in Parisian neighborhoods in 2004, 53,391 of which were pulped, around 10,000 put in storage, and the rest ending up as charitable donations. The bookstores with bursting bladders dream every night that they are sending back cases of new books unopened, and some have obviously passed from wish into action, while depression and renunciation take over the bookstores of the whole world: no room, too many books, not enough readers.

It is an impasse whose very entrance is becoming cluttered. Only the transformation into "bits" can inspire hope for a much-needed decompression that still has the power to shock today but can no longer be avoided: leafing through books online and having them delivered, manufacturing out-of-print books one copy at a time, reading or working entirely with virtual content—whether it costs a fee or not—all gestures that soon become banal following the most amazing innovations. May the keepers of the network* be reassured: the old market of the perishable and the four seasons will endure for yet a little more and continue paying the shareholders long enough for them to seek new or replacement capital.

Reading on screen is difficult? But ruining your eyes on a "microform" is much worse, all the professionals of forced decoding will retort. A large, noisy, and dirty machine—one generally already in use when needed—is necessary to enlarge and zigzag the pages about in the vain effort to explore them. I have already described these pitfalls, as well as the despicable reasons that are behind their widespread use, but the real tragedy lies in the fact that once a book has been transferred onto plastic, it ceases to be circulated in its original form, even if it is still in fairly good condition. How many books like this are currently unjustly experiencing the enduring stillness of the tomb? I have lost count of the books that the BnF has forced me to go consult at the Sorbonne or the Institut. What is the world ranking that a National Library descends to when only the books that are used are counted, meaning those in paper and free to be consumed without restraint?

> I had to read one hundred fifty e-books on my PDA. . . . And I was not the only one; for example, in October 2005, sixty thousand books in PDA format were downloaded on the site. . . . If the PDA is good, reading is quite comfortable. When it comes

*[The play on words the author has employed here does not exist in English. The French word *filière* means both network and wire drawplate, among other things. —*Trans.*] The wire drawplate, created in 1829, is a machine that pulls metal into thinner and thinner threads. The network/drawplate of the book is, by extension, this system into which the author enters with all his abilities (his language, his enthusiasm, his logic, his dreams) and from which the final product emerges at the other end.

> to the ergonomics, with software like Mobipocket, for example, I found it superior to a paper book: one can mark the pages, take notes, underline, and easily navigate with the table of contents, and so forth. Personally I did not experience any visual fatigue. The important thing is to choose a PDA that has a nice luminous screen with sufficient contrast.

This testimony, signed Coolmicro (and amended for the present quotation*), can be found on the free Ebooks site whose address appears on the back. Similarly, by using the services of an outside company to provide content and online accommodation,† the Landowski Library of Boulogne-Billancourt[2] offers its members a thousand time-degradable (changed) books, some 80 percent of literature that Landowski makes available in different languages, which people can scroll through on their pocket machine or computer, and undoubtedly soon on the new Japanese tablets, without having to travel anywhere. This winning hand has permitted the establishment of a BNH (a digitalized library for the

*The author adds: "The border between the PDA and the computer is shrinking, even disappearing, with the small portable computers weighing less than three pounds or the tablet PC. The number of people using these small computers for reading, both with software designed for the PDA and with Adobe Acrobat, is increasing, and it should be recognized that these machines are extremely pleasant for home use. There are also the machines specifically intended for reading e-books like the Bookeen Cybook. All these machines are going to evolve substantially during the next few years, the number and variety of terminals will increase, and these terminals will become more mobile, easier to handle, less expensive, and so forth. Therefore one can picture that there will be several kinds of 'tablet-screens' of various sizes in the home, with a wireless connection to the main computer on which anything can be done such as surf the Internet as well as read electronic books. Furthermore, the appearance of flexible 'pagelike' display screens (already existing as prototypes) will make it possible to come close to duplicating the sensation of a paper book" (e-mail of 11/21/05).

†This new kind of provider is known as an "aggregator." It is responsible for both the digitization of the pages and publisher relations, thereby relieving the library of various concerns, notably the question of copyright for the loaned book. For the member, it is still free (if he or she is from Boulogne, in this instance; otherwise a modest annual fee is imposed on the user).

handicapped) managed by Alain Patez, which began as a special service intended for people suffering from lissencephaly (brain disorder), quadriplegia, and muteness.

If one adds all the other advantages to charity, such as perpetual access to collections, the automatic restitution of the borrowed book, the disappearance of the need for inventories, and the dread of deteriorating stock, not to mention a vastly less expensive purchase price than required for a paper copy, how could any librarian with a pioneering soul, or administrator for that matter, hesitate before such a clear future? But here again, the ranks of the publishing world most set in their ways are applying the brakes to this movement, convinced that a reader of Wittgenstein on the screen is concealing a troubled personality, truly a vicious predator.

Jean Gattégno joined Ferdinand Foch and Darryl Zanuck in the circle of dumbfounded prophets the day he declared Proust could not be read on the screen.* However, a certain Lewis Carroll–like fluidity may have allowed him to conceive that our eyes are changing, and those of the next generation even more, and that computer screens are helping push this process along. (Already novels of all sizes are being devoured on a daily basis on mobile phones from the comfort of one's *home* in Chinese or Japanese, whose writing obviously facilitates the phenomenon.) Furthermore, the pages of books online are at risk of being transformed from the dry reproduction of a typographical block—invented at the same time as paper in order to imitate the manuscript in every way—into a mobile and flexible entity we can still only vaguely imagine based on various experiments here and there. There is the old TextArc, for example, which creates visual promenades of *Alice in Wonderland*

*"Airplanes are interesting scientific toys, but offer no military value"—Marshall Ferdinand Foch, 1911. "Television won't last because people will soon get tired of staring at a plywood box every night"—Darryl Zanuck, 1946. But it is especially curious, with respect to Jean Gattégno's prediction, that the similar declarations from other countries, even the non-French-speaking ones, all provided Proust as their benchmark. There is much, no doubt, that this observation could teach us.

or *Swann's Way,* to be specific. We then have the software that softens codex pages like the Turning the Pages program the British Library offers visitors to its online site for perusing some of the gems of its collection. Also under close watch is the research performed by Pierre Cubaud of the Cedric team of the CNAM (Conservatoire National des Arts et Métiers), which has a parallel bearing on the dynamic transformation of three-dimensional catalogs.[3] On the other hand, electronic paper, or *e-paper,* will not be long in finding applications. Its prototypes can be scrolled like papyrus today, and e-paper will be made available in color tomorrow. Progress will be equally rapid in solving the problem of luminosity, which at present is too weak and too tiring on the eyes, or that of pixilization. Also impatiently awaited is the advent of a fine-tuned modular movement system for scrolling smoothly through the lines instead of suffering through the intolerable hiccups of Adobe Reader, which is so crude it cannot help but improve. The mechanical stiffness of the text can be compared only to that of the newborn robot.

A deep breath filling the lungs suddenly propelled study and research forward with their rapid electronic fluttering of content from one screen to the next, simultaneous postings, transfers, and so forth. The inflow of texts palpitating online is a relief to all those who are forced to come up with an essay in six months or give a conference next Sunday morning. These individuals all know that, in principle, the author of an essay does not have to appear inordinately intelligent or educated. It has always been the responsibility of his or her reader either to possess those qualities already or to gain them, thanks to the author. But another question keeps popping up: Does the fact that the student has unearthed the book through nontraditional methods prevent him from reading it in its entirety? The more easily upset individuals fear that the new form of reading will only reduce the individual's ability to reflect. Instead of benefiting from the slow digestion of a book that haunts and utterly transforms him—people do commit suicide, you remind yourself, on closing *The Sufferings of Young Werther*—the reader, out of laziness, will become acclimated to pointillist thinking in the form of digests and snippets that can only scratch consciousness at best, then unhesitatingly adapt to this mutating text that consists of vignettes and perpetual appetizers to which the eye's new habits will next give birth.

6

A Digital Coronary

Chronologically speaking, the first victim of Google was not Jean-Noël Jeanneney but Michael Gorman, who, stripping bare his unconscious chest, suddenly exposed the omnipresence of the future's execution squads to the light.[1]

Dean of Library Services at California State Fresno, and more importantly president-elect at the time of the ALA, the immense association of American librarians, Michael Gorman, who is sixty-five but like all good media-savvy Californians looks twenty years younger, published an op-ed in the December 17, 2004, *Los Angeles Times* on the "boogie-woogie Google boys" who were mounting an assault on library contents to obtain an exhaustive collection of knowledge that, in their opinion, was comparable to "the mind of God." What ignorance, argues Gorman. A book is not just the sum of its parts, and any researcher worthy of the name should always read it in its entirety, or simply be satisfied with pecking information out of it, which is not knowledge, hence the title of his opinion piece: *Google and God's Mind: The Problem Is, Information Isn't Knowledge.* Digitization is useful, he adds—truly, it is essential—for those books in which information is accumulated without a context like encyclopedias, or even for very valuable manuscripts, or photographs, and so on (this was reportedly the deal made by the New York Public Library with Google). He concludes that plans for massive databases of books online, especially scholarly works, "are expensive exercises in futility based on the staggering notion that, for the first time in history, one form of communication (electronic) will supplant and obliterate all previous forms. . . . This latest version of Google hype will no doubt join taking personal commuter helicopters and carrying the Library of Congress in a briefcase on microfilm as 'back to the future' failures, for the simple reason that they were solutions in search of a problem." Fans of aporia will appreciate what he said, which does not prevent this fine man from being both completely right and completely wrong—something of which someone of his eminent position should have been vaguely aware.*

That some hidebound relic from the old school periodically lashes out with a reactionary article to titillate the younger generation, and to see what their insides are made of, is simply a fact of life. Generally, the author of such screeds will lay on the provocative notes quite heavily, and then will slip in a pinch of self-mockery at the time the dish is served. But what was commonplace in classic intellectual debate is apparently no longer possible today. Michael Gorman is well positioned to know that no one writes anything today unless his or her lawyer is present.

He did not have to wait until the next day for the storm of criticism to descend. This John Wayne of librarians, surprised by the sheer number of comments he elicited, which were quite belligerent and even came in on his own laptop, immediately plunged back into the melee with a new article in which Google, "a wonderfully modern manifestation of the triumph of hope and boosterism over reality,"

*A response could have been made to his main argument, but no one in the United States seems to have taken it seriously—that the sole difference a paper book carries—in addition to the clearly superior epidermal pleasure it provides over that produced by touching plastic (but nothing prevents us, after all, from covering them in vellum) is that the total weight of the text is constantly felt by the reader. This sensation perhaps gives the reader an impression bordering on fetishism of possessing the whole of its meaning, an illusion whose loss could panic fragile souls. This is because my left hand is holding the beginning of the work and my right hand the end; and my gaze knows precisely where the idea or point in the story is situated between the two.

was quickly set aside to mount an attack against its alleged defenders, the bloggers who, possessed of "a fanatical belief in the transforming power of digitization . . . judge me to be wrong on the basis of what they think rather than what I actually wrote. . . . I doubt that many of the Blog People are in the habit of sustained reading of complex texts," and so on and so forth. As the sycophants of the Web are faceless, a torrent of verbal violence welcomed these remarks: more than a thousand reactions on the technoid site Slashdot and at least one hundred in Gorman's office, only one of which was approving. Finally, his resignation from all his duties was announced by someone and echoed across the Web within an hour by surfers worldwide, but this took place on April 1.

The moral of this fable? First, electronic chewing gum has a strong tendency to render vague even writing of absolute clarity, while the quasi-general anonymity of the new form of correspondence stirs up doubt, insult, incrimination, and April Fool's gags. Second, an entire population that feel high-tech wings sprouting from their backs seem to stuff a priori into the tackiness basket everything they do not (strictly) know (yet). Erudition is thus called intellectualism, or the librarian is deemed "anti-techno," whereas, quite to the contrary, there were already 326 blogs by English librarians at the beginning of 2004, all quite up to date on the latest innovations of information technology, sometimes to the point of appearing to be in tune with even the most abstruse developments. If, as we are told, we are witnessing the emergence of a countervailing power, it will soon be high time to forcibly introduce into it a few spoonfuls of intuition, humor, and relativity.*

*Only one single book, somewhat profound although entertaining to read, seems to exist that deals with—among numerous other observations on the midterm and long-term obsolescence of companies, mass media, governments, libraries, universities, and so forth—the absence of distance that the dematerialization of daily life generates and aggravates. *The Social Life of Information* sheds light on the fact that with everything becoming information, we shall soon have nothing left to talk about but information. For the moment, no French publishers seem to have gotten wind of the burning pertinence of this work, now translated into several languages, which was first published in 2000 (John Seely Brown and Paul Daguid, *The Social Life of Information,* Boston: Harvard Business School Press, 2002).

7

When the Book Is Too Highly Concentrated, the Purpose It Serves Is Easily Forgotten

There was a time, one that lasted for centuries and is falling into the *oubliettes,* when any publisher worthy of the name would take all kinds of textual risks and balance his bottom line with one or two successes over the year. Today, the purchaser of a book can hardly be unaware that each title, for all of the large houses, represents a *profit center,* an enterprise in and of itself that undermines, even before the first line has been written, a percentage of the profits it will provide within a limited time frame. Everyone knows it, everyone deplores it, and everyone does it. One of the most powerful American publishers even has the courtesy to warn future geniuses that it is useless to send even the tiniest manuscript: no book will appear, says Knopf, that is not written in accordance with a list of requirements put together by the directors of our catalog, as they alone have perfect knowledge of the market and its current demands. Furthermore, publishers are so aware of the market that they are perpetually prepared to swallow one another. This process of the despoiling and cannibalizing of publishing houses in the United States is clearly described by André Schiffin in a spurt of two pamphlets (Eric Hazan, who published them in France, was himself a victim of this kind of sinister event, as he recounts in the magazine *Esprit*[1]). For example in London, in July 2004, the very serious Frank Cass and Company eventually surrendered to the voracious Taylor & Francis group after a half-century of independence because its high reputation as a scientific publisher no longer sufficed to induce superstores to allot shelf space to its publications, or was actually a handicap. This is the rule of "cultural censorship," the Cass shareholders fretted and fumed, as they pocketed fifteen million in pounds sterling. The situation of the bookstore in the United Kingdom is far more disastrous than in other European countries, but, as is the case with the cost of housing, with the general deterioration of the daily newspapers and travelers bombed in the Tube, "all evil comes from the north."[2]

And this is why the book is inconsequential. This phenomenon that has made the cashier the decisive influence over the text ensures that soon you will have to choose every year between, at the very most, twelve titles that either copy each other or crib from the dead: Tolkien in his time made the individuals who had read a few tales burst out laughing in commiseration, and now he has become the bottomless well for lucrative inspirations.

Let's remain magnanimous: three-fourths of what is published does not deserve to be. At least, this is the case if one sticks to rational and honorable publishing criteria such as originality of subject, style, and ideas. The obvious inference is that publishing can no longer use rational and honorable criteria if it has to satisfy the desiderata of management's auditors, as all the studies cited declare to a nauseating degree.

Here, as allegedly quick-witted and honest readers who have grown aware of the raft rotting beneath our feet, let's take a look together at what we are losing. With the definitive absence from bookstores of the millions of wonderful books that have been written, printed, and sold, and for a long time patiently republished, the publisher-distributors of each territory—perhaps soon to be shrunk to the number of two—do not have the ability, the knowledge, or the desire to offer them to you. To do this, they tell you, it would be necessary to do a three-hundred-copy print run of the books that you truly wish to discover (how capricious!), without any marketing expense—for example, because you would have read at least an entire page. And, what's more, you want them home-delivered? What an aberration! It is not at all economical! This would also be, yelps the government representative in turn, the death of the proper channels!

Then out of the night looms a new kind of knight of industry.

The names are Amazon, Google, and soon other names of the same more or less infantile scale. Are they coming to shake the foundations of antique buildings? They are barely aware that such things even exist. They are irresistibly pushing their virtual plow, thanks to the original and often cursory way they envision communications, law, and commerce. And the dust—somewhat similar to fallout—stirred up by their passage shows that they are not the only ones active in the field of digitized books. Here and there we find numerous isolated libertarian practices, cooperative activities that also casually implement the gifts of the technological revolution. All are headed toward the same goal of a boundless diffusion, despite their barely compatible philosophies, and they will soon cause the people in the "trade" to confront what they most dread: the necessity to question the very foundations of their profession, if not simply to disappear.

Before the library project of Google Print had even been announced, the online bookseller Amazon had already discreetly digitized a substantial portion of the books in its English catalog to offer the snoop the possibility of browsing through a few whole pages. The operation consists of photographing ("scanning," rather), then analyzing an entire book with a character-recognition system in such a way that search robots can detect each of its constituent phrases. Any curious individual seated in front of a keyboard will theoretically be able to find any book by invoking any phrase. Until now, only the best paper documents and essays were embellished with an index designating the page on which the desired name or concept could be found. There even existed thesauruses, analyses by key words . . . A plague on these industrious bad habits! Mrs. Modernity seems to think that today, every word of a book can be contained in an absolute index. The notion can thus be put forth that the index is the book itself. This is somewhat miraculous and incontestably useful, but in the eyes of those deriving income from the former conditions, it is a proposition for rape.

Here is the record of an experiment: you wish to verify how Horace Walpole forged the neologism *serendipity.* On Amazon, this word triggers the appearance of the book *Serendipity: Accidental Discoveries in Science,* Wiley Edition, 1989, within which you are invited to search. You input the search term "Walpole." Immediately a list of the phrases concerning him appears. You can then open several pages for polite reading, two pages after and two pages before. Note down the information required and gratefully close it all back up. Whether you buy the book depends on the level of its quality compared to that of your requirements.

This system, baptized "Search inside the book,"* already accompanied 120,000 titles offered by Amazon in 2003. The journalist Gary Wolf recounts in an article in the magazine *Wired* how, when researching a con man named Boss Tweed, he typed this name into the Amazon search engine. Not surprisingly, the machine immediately started churning out books devoted to this figure, but also unexpected titles in which he was cited, even for a slim and perhaps promising detail. These included *A Confederacy of Dunces* by John Kennedy Toole, *American Psycho* by Bret Easton Ellis, and *Forever: A Novel* by Pete Hamill. It is undeniable that no synthetic analysis of reading material from the old days would have had the slightest objective reason to bring out this kind of detail in the strict framework of a list of operating instructions.[3]

Digitization already allowed us to judge the seriousness of a book by

*[And now simply Search Inside! —*Ed.*]

its table of contents or its *priere d'inserer,** as well as of its style through a pair of pages. Today it is the whole of a text that is laid out flat.

In French, "Search inside the book" became the awkward "Chercher au Coeur!" [Search Inside]. Now or never is the time to type *alfacinhas* and discover how the Lisboans, formerly called Lisbonnins and Lisbonnines, were nicknamed "little salads." This is on page 2 of the *Petit Futé, Lisbonne 2005–2006.* You get to read the two pages before and two pages after. The publishers Dargaud, Ellipses, Belles-Lettres, Trajectoire, and Klincksieck have already made themselves over into Amazonians, if not Amazonists, in the footsteps of the pioneer La Découverte. In 2004, it became obvious to François Gèze, the director of this publishing house, that online browsing was merely an adaptation of what took place in a bookstore. The 7 percent of hoped-for additional sales would not, as the vast majority of the French publishing community dreaded, be outweighed by any hidden danger. The thinking at La Découverte was that once those works whose purchase value would be nullified by their online availability had been removed, nine hundred books remained available in the backlist, which this long-distance bookseller briskly digitized and put on its site in the middle of 2005. Is there any *thigh* in *Les Débuts du juge Ti* [The Debuts of Judge Ti]? Yes, it appears twice, once scratched and once knocked. And can any *soul* be found there? Yes, in thirteen passages. Are there any *souls* in *Les Bibliothèques* [The Libraries] by Anne-Marie Bertrand? Not a one, responds the stupid search engine, which has bizarrely skipped over the one on page 89 where municipal libraries will supply their communities with an additional bit of soul, as the author pertinently assures us. In compensation, if I may put it that way, this book mentions *rap* on three occasions, but not a single *thigh,* even on the BPI.†

Once under way, could the cunning freeloader take advantage of the

*[*Prière d'inserer,* "Please insert," is a one-page or longer insert that describes a book's features, much as the back cover and jacket flaps do for English-language books. —*Trans.*]
†[Public Information Library. —*Trans.*]

system to use key words like *but/or/nor/and/therefore/thus/so/because/as/the* to gain access to almost all the pages of a book? None of these words finds a corresponding "reference in this book" is the response that greets the user sly enough to try one, which exposes a debatable selection of key words at the source and corroborates what the erasure of the *soul* in Anne-Marie Bertrand's book led one to suspect. On the other hand, the same open sesame works perfectly well to obtain access to the cavern of Google Print in French, which plays the game of perfect indexation, although it forbids reading of the entire text by locking away certain pages and imposing an automatic quota of pages on each reader. Why? such a reader might dare retort, rebelliously. Because, answers the robot—who, although he cannot think, is at least well meaning—I am not here to give you books to read, but to help you find them (and judge them, I might add).

Amazon's April 4, 2005, acquisition of BookSurge—a service specializing in, among other things, the printing on demand of out-of-print books—is clear confirmation of a realization and its unavowed objective: selling merchandise is more profitable when it is rare and costs the seller hardly anything except for the reversion of a few cents to the author, if he is alive and if he shows up to claim them. While awaiting this push toward collections that are more out-of-print than in fashion, BookSurge is already offering services, keys in hand, to every level of the publishing chain, to bookstores as well as to the apprentice author enticed by the Amazonian pennant, including such assistance as writing the book for him or her and printing twenty-five copies at fifteen dollars apiece. Free shipping for fifty or more copies.

As we shall see later, giving substance back to ghosts is well on the way to becoming a commonplace gesture. Thanks to a gift from Steven Spielberg, the Yiddish Book Center of Aaron Lansky, nicknamed the "Noah of Yiddish,"[4] has digitized a portion of its 1.5 million books, collected from all over the world.[5] A copy printed on demand from this digitized copy will cost the would-be reader forty-eight dollars. The price seems high for a paperback book, but the book is almost a *unicum*.

The advertising presumptuousness of Amazon.com was once the target of mockery: of the hundreds of thousands of titles that Jeff Bezos boasted he could put online, 80 percent were in fact unavailable from their publishers. But candor was not on the side you might assume. Keep your eyes on what he's up to, one might suggest: As the crowning achievement that gave this multi-tentacled and Babel-like edifice its full significance, on March 30, 2005, this company also bought Mobipocket, the Parisian publisher of e-books and the digitizing software of the same name.

He who shines flamboyantly at his stand at the London Book Fair shall weep when scraping the aisles at the Salon du Livre in Paris. The large British publishing houses in 2005 gave a chilly reception to the Google Print proposition, even taking umbrage at the fact that the publishing house's union had invited Google Print to exhibit, and spoke of *Napsterization,* a neologism forged from Napster, the now vanished pioneer of the accursed music downloading sites. The publisher of Harry Potter books went so far as to prophesy kabbalistically that while acquainting readers with books via digitized extracts would swell sales for the short term, it would bring about their collapse in twenty years. The reason for this eternal span of time is unknown, but the profession trembled, despite the fact that the professional periodical *The Bookseller* proclaimed quite the opposite, saying Google Print was "one of the most promising opportunities for publishers."[6] Thus, at the Porte de Versaille* some fifteen days later, the Google representative sought to form contacts while going from one stand to the next—with a stack of documents clasped to his chest, so that his badge would not be the first thing his interlocutors saw. Only one lone exhibitor gave him a warm welcome, the one who already had personal experience with books being available online—to be read at no cost—at the same time as these very books were being sold in bookstores.

Michel Valensi created Éditions de l'Éclat in 1985 with the first

*[Site of the Salon du Livre. —*Trans.*]

volume of a collection called "Imaginary Philosophy" that in 1993 would become home for a bilingual version of Pico della Mirandola's *Oratory on the Dignity of Man,* with a 2,500-copy print run. When no more than a hundred copies remained in stock, the publisher had the fantasy of putting the entire text online on his site, as this is how things are done in the tranquil lives of public domain and independent publishers. The demand in the bookstore became such that the book was quickly in its fifth 2,000-copy print run. Pico della Mirandola, blasé precursor to all things modern, was followed by a gift to the Internauts by the same publisher of several works that the Internauts/readers are also authorized to purchase in the more standard manner. This is something the readers have not failed to do, buying such books as the collection of foundational essays *Libres enfants du savoir numérique, un anthologue du "libre"* [Free Children of digitized knowledge, an anthology of the "free/book"], edited by Olivier Blondeau and Florent Latrive (March 2000). This house was the first to reach accommodation with the revolutionary tool offered by Google, and 163 of its titles made the trip from Paris to Mountain View at the end of September 2005 to conquer the planet, perhaps, by making themselves available for virtual browsing. The first thus served and thus honored: Spinoza's *Ethics,* translated with commentary by Robert Misrahi; and *Déclinaisons de l'Europe* [Declinations of Europe] by Massimo Cacciari, a former mayor of Venice who maintained the European Union would have to go through a profound decline before beginning to blossom. Four weeks after passing the first ninety-nine titles through the huge Google mill, the publisher read his first activity report: 45,715 pages consulted and 440 clicks on the Buy button; particularly difficult titles selling three copies a month went up to fourteen . . .

So which should it be, Amazon Search Inside! or Google Print? For the moment, their only differences seem superficial. The first merely transposes the bookstore display window before which the Internaut stops after he or she has entered the Amazon commercial gallery, having already been led there with a book in mind. The other proclaims its desire to put any and every book into orbit in the hierarchy-less scrum of Web pages. Here it can be stumbled upon by a fan or researcher hunting information that he or she did not necessarily think would be found in the form of a book, and it is undeniable that the work benefits here from potentially being pushed into an infinitely more powerful spotlight than that provided by any other tool. As an American critic recently said with perhaps just a dash of irony in his choice of topic, if a new, extremely erudite essay were to be published anywhere about the Maginot Line, the three hundred individuals in the United States who take a close interest in the Maginot Line would quickly become aware that it was available. A result like this can be achieved only with a disproportionate publicity effort, thus one that would have been unthinkable just a short time ago. Literature sees the dawning here of a generous collateral benefit: the possibility of publishing and even printing impossible books; no matter how contorted the novel or abstruse the theory, there will necessarily be a kindred soul somewhere.

In the first case, we see the traditional merchant trampled down, soon to be supplanted by the modernization of the tools of the grocery store, which is as old as humanity. In the other case we see, in addition to the surgical precision it employs to connect the book to the one who wants it, the possible conquest of as-yet-unknown buyers who do not dream of its existence. The venture is charging ahead.

Reading online is thus sketched out just beneath the surface in all the promotional good intentions of literary merchandise, titillating great operators and medium-size publishers as well as minor authors. Without having to read the future, it is already possible to follow the strategic maneuvering and the buyouts of complementary companies that owe nothing to chance. Furthermore, Amazon Shorts is beginning to offer short, unpublished texts (less than 10,000 words) for forty-nine cents. The reader chooses the format (full text, PDF, or HTML) and becomes owner of the file, which he can read, transmit, or print. On August 19, 2005, the day Amazon Shorts was launched, sixty-four titles were quietly folded into this company's general catalog. The writer retained in this manner grants Amazon exclusive rights for six months, and in return is offered a very honest gold mine: a 40 percent royalty.

So here Amazon has suddenly become a publisher. At this juncture, who finds this surprising?

8

The Pixel Coming to Paper's Aid

Stephen King halted his page-by-page delivery of *The Plant,* a novel written for the Web, to the great dismay of all those who had paid for each of the first five chapters as they came out (a total of seven dollars). Thus Stephen King did not follow through on his promised work (but just you try and institute a lawsuit against him). The experiment tempted him, but its outcome was disappointing, this disdainful figure said. But because of his action, other American writers maintained their interest in electronic self-publishing.

After his manuscript has inspired the snickering of every agent, and thus every publisher, more than one mule-headed author has stuck it not in his basement but on his blog. Miracles sometimes happen: hits explode, people start asking for it, and agents, who are shameless, start knocking on the door. *Wired* magazine cites as examples Brandon Massey (*Thunderland*), Victoria Christopher Murray (*Temptation*), and Travis Hunter (*The Hearts of Men*), all works that had been offered in their entirety on the Web and which were extended by large paper printings.[1] One assumes the publishers were made to pay dearly.

NAP (National Academies Press) is an American publishing house that was created under the aegis of the National Academy of Sciences and related institutions and operates under a charter granted by Congress. Because the content of some of the two hundred books a year it publishes is deemed vital, it offers the ability to read them page by page online for free, at the same time it makes it possible for the reader to order a hard copy at any moment. The house has been doing this, it proudly announces, since 1996.[2]

We do not know what results one Laurent Chemla obtained for his book *Confessions d'un voleur: Internet la liberté confisquée* [Confessions of a Thief: The Internet, Confiscated Freedom] (Denoël, 2002). This author, quite concerned about what was taking place in the wings of the Internet, was the first (in France) to get his publisher to agree to distributing his book to bookstores at the same time he made his manuscript freely accessible on the Web. Florent Latrive, who made the same arrangement for his *Du bon usage . . .*, did not respond to questions concerning the comparative outcomes of these two experiments. Luckily we have Korea. The book *49 Things to Do in Your Life*—keep a diary and read a book are some of the occupations readers are strongly advised to practice at least fifteen minutes a day—published by Wisdom House, sold 800,000 copies in 2004. It was then put online for less than half the cover price, which produced a much higher profit, given that the costs (not including royalties) were close to zero. This book was downloaded fifty thousand times in 2005 and continued its brilliant success in the bookstores.

In France, for a traditional book costing 10 euros at the bookstore, "the breakdown is as follows: 0.55 euro for the value added tax, 1 euro for the author, 3.8 euros for the bookstore, 1.5 euros for the wholesaler/distributor, and the 3.5 euros that remains are for the publisher. The publisher figures 2 euros for manufacturing costs that, in the event of a reprint, falls to 1 euro."[3] Both good and bad writers are therefore aware that some 90 percent of their sales figure is sprinkled over the head of a large household, who, by and large, deserves it, it is true—most of the time. More than one is satisfied with this munificence to live the life of a maharajah. Others are perhaps going to feel the extent to which the cybernetic revolution has also changed the hand dealt for their activity, even if their attitude is hostile or scornful for the most part. At this time, they are still at risk of learning the new rules of the game—truly, if need be, creating their own.

9

Is This Already the Post-Google Era?

Steering a flawless course is the easiest thing in the world once you have placed yourself in a position where you can live on the work of others. For example, sales-convention organizers will tell you: simply pick a theme and a good date; all that is left for you to do is cash the checks of the exhibitors and enjoy a gross return of 1,000 percent. With search engines on the Web, the profits are even higher, as there is no physical labor required: robots scour and rank existing pages in accordance with their notoriety, surfers and surfed contribute unawares all their nights of sweat and toil while your sales staff are selling advertising spaces on commission. Just a few short years after start-up, basking in the world's gratitude for having provided a free service by having it funded through other means, your company is placed on the stock exchange, and here you are among the hundred richest individuals of that moment. So how will you keep busy now?

Underneath the limpet lurks the octopus. During the year 2004, Google Incorporated began spreading its tentacles in all directions; it went unnoticed for some time how nimbly they came back together after having fished in many waters:

- *Publishing:* Under the trade name Google Print, it offers book publishers its resources to publish extracts, notices, their catalogs, all designed to induce the user to order the book online from one of the specialists in long-distance selling or from a local bookseller.
- *Scientific research:* In a way that is both free and confidential, it is publishing a jungle of articles that are of the utmost importance for new research—finding these articles and then finding them again is facilitated by Google Scholar.
- *Large libraries:* The revolution of learning promised by science fiction is scheduled for tomorrow. This is a topic we will revisit in depth later.*
- *Television, telephone, news programs, maps, and so forth* are also offered under what are or soon will be appropriate names (Google Talk, Google Video, Google News, Google Earth . . .).
- *Consumerism* is facilitated by a kind of shopping magazine that has been ironically (or diabolically) baptized Froogle, pronounced like the adjective meaning "thrifty": frugal.
- *Professionals* who already have a well-established Web presence, thanks to Google's tempting offer to add to their site the free and splendid tool of its search engine, which is now the tool of choice by an overwhelming margin, are now offered a bonus. Google will send you a nice check at the end of the year (for example, all the major libraries like the BnF have thereby introduced the worm into their poorly defended fruit without imagining that the number of "hits" figures on which their remuneration is based also serve their service provider as studies for its surprise projects).
- *Everyone's personal computer* is enhanced by equipping it with

*Some of the details in this case file have not been summarized, as they should be known to all thanks to the large media coverage they inspired. There is, however, a large section on them in appendices 1 and 2.

> Google Desktop, which internally indexes the shambles our machines have become—Web bookmarks, or our "favorites," all the various kinds of mail, and text files, not to mention the images and sounds astride them. This mass of material is constantly doubling in size, and any given item becomes so hard to unearth that no one has the slightest hesitation about handing over the keys to their home to the first comer, provided he does the housework. It goes without saying that this silent factotum will take advantage of this to keep your IP address and all kinds of information that over the long term will create a file worth billions, capital someone will know just how to use one day, like the recording of *all* the clicks and keywords made to this day on the search engine (and never erased). This is the most formidable and least burdensome market study ever conceived. It is also, inveigh the Cassandras, who are certainly in the right here, the most colossal human entanglement of all time.

The fantastic menu of services will obviously lengthen each morning, to the point that the list will be consulted while you drink your coffee in the same way you study the stock exchange pages. All that is required is a visit to the site of the motherhouse, where, after a detour through the premises of "Googleplex" and its prodigious output, guaranteeing its personnel instant obesity, the tab "Labs" demonstrates what resources are now being hatched. Holding their hands over their hearts, your hosts, Mr. Brin and Mr. Page, swear they are only trying to do the planet a favor. And, on the whole, who among us would not react like Disney's Pinocchio to the barley-sugar forest?

So is there a hidden face of Google, and is it black, and is it the size of an iceberg?

This new Uncle Scrooge, once it had counted the three billion dollars available in its treasury, announced in August 2005 the sale of new shares expected to return an additional four billion. Of course, the exchange bubble can always burst at any moment, but the enormous profits hoped for were by and large more than met.*

And what did the company do with this small fortune? As the first means of economizing is limiting means, it bought up bankrupted telecommunications companies or parts of wireless sectors by the dozens, in order to arrange conduits that would serve it well (the user pays his service provider for access, but Google has the responsibility of transmitting content to the latter from its own servers), and realized a huge profit. But even this seems negligible in relation to the use this house bandwidth could permit: the creation of global television service, landline and mobile telephones, and all other services both free and fee-based, existing and yet to be invented, that for the most part are irresistible. The company has already offered to cover San Francisco with free Internet access, both wireless and with no strings attached, except for publicity. These observations have given rise to the hypothesis that after it has crushed all the other search engines and has become everyone's connection to the Web, then Google *will be* the Internet. All that will remain for the company to do is create a new operating system to get under Microsoft's skin as a way to amuse itself. "By this token, if the Attila of search engines persists in pushing its myriad of projects forward, it will set the stage, beyond any doubt, for a Google-type confrontation with the rest of the world."[1] As many observers have emphasized, this future GoogleNet will hold each of its users in its net in order to perpetually force-feed it personal advice with lurking advertising, to paint only one of the rosier scenarios. It will not be Big Brother. It will be Enormous Brother.

---

*The firm, which achieved a $3.2 billion figure in sales in 2004, multiplied its profit by seven in each trimester of 2005 in relation to the preceding trimester. The price of a share, introduced on the exchange in August 2004 at $85, went over $300 ten months later. The capital provided by the exchange was then more than $87 billion, a figure that outpaced not only Yahoo but Disney and Time Warner as well (Nasdaq.com, July 2005). It continued to grow: $1.4 billion for the second trimester of 2005, $1.58 billion the following one. When the cost of a share passed $460, some analysts advised selling while others "felt" it would climb to $600. [As of August 2007, Google shares were trading around $525, after reaching a high of over $700 a share. As of October 5, 2008, Google's stock was down to $371/share. —*Trans.*]

10

But Why the Devil Do We Need Libraries?

Profit comes more freely to someone emancipated from any physical presence. While the industrious Bill Gates continues putting out proprietary software (Africa is the only remaining spot in the world for this upstanding citizen to impose his Windows, but thanks to the benevolence of UNESCO, he is well on the way to achieving this), the companies climb higher if they offer less, or at the most an algorithm, in other words, the algebraic wind that is swelling the sails of all this activity.

This unyielding principle implies that it will not be long before the firm of Mountain View is swept away by one that is yet more volatile. To outpace the competition, something solid to put into the networks is required: content. A lot of content and content, which costs nothing . . .

A computer-science genius being, by definition, completely devoid of creativity, his gaze falls naturally upon what is already before his eyes: a doughnut, a can of soda, and, beyond them, books. The founders of Google are a couple of aging students. Because their experiments and their company originated in their observation of the difficulty of sorting through data at college,* it was their university libraries of Stanford

*A rough draft of the epic was announced in February 2004 under the name Project Ocean, but the name was obviously dropped with great eagerness, no doubt because it was too evocative of a mass drowning.

and Michigan that were first to get laid—an expression that might seem excessive, but which the vision of the process inevitably brings to mind. Huge, metal, entirely automatic machines scan the books by turning the pages at a quick pace and can tell—based on weight—if they have skipped one, in which case they start going backward. Another robotic process analyzes the letters and simultaneously transforms the book into a completely indexed file that the user can send and manipulate.

A slight churning of the stomach accompanies examination of the figures: fifteen million works, to borrow the media's leitmotif, equals close to five billion pages to be turned, OCR scanned, and corrected, then cataloged and stored in giant servers to wait for some customer to indolently rummage through them. A battalion of Hecatonchires* would not be enough.

The sole thing left unknown is for how long Google has been providing assistance with its advice, orders, and financing of tests to the rare manufacturers of machines that would permit imagining astronomical quantities of books being digitized at reduced cost in both time and money. If their robots are truly capable of turning pages—"gently," as their press releases claim—at a rate of 1,000 to 1,500 an hour, working nonstop day and night, that would theoretically give us at least eighty digitized books a day and thirty thousand a year. Based on this rhythm, it would take a hundred units divvied up among the largest libraries to reach fifteen million books in five years, if it is truly necessary to achieve the impossible, which is not an absolute obligation. To avoid weighing down these pages with technical details that will bore some readers and, given the time it takes for a book to be produced and printed, strike others as vastly out of date, links to companies claiming to be experienced in this kind of mechanization are supplied in an endnote.[1] One of these companies was already well into the swing of things in Stanford before the announcement that caused such alarm in the press in December 2004—it perhaps even inspired the idea—and

*[The Hecatonchires are hundred-handed giants in Greek mythology, the children of Uranus and Gaia. —*Trans.*]

in all likelihood is still slogging away there, although the spokesperson does not respond to any questions on this matter and ferociously keeps such intrusions at a distance. Another company attempted to seduce the French decision makers at the end of 2005, while the Internet Archive rather lightheartedly admitted it had perfected a machine in Toronto nicknamed Scribe that simultaneously turned the recto and verso pages manually, for a price five times less and at a speed theoretically doubled. But here as elsewhere, while all this technical transition now seems somewhat laborious, in the near future it will be a widely used routine.

In the meantime, this is perhaps where destiny lies waiting for the search engine–turned–content provider, at the detour of the physical labor of the massive digitization and its impedimenta, the T-shirt soaking wet from moving tons of often fragile books for a result that will no doubt be studded with misprints to be touched up one by one against the backdrop of haughty heirs lying in ambush, and so forth. Once the company has exhausted itself moving boxes and is mired in court actions for the foreseeable future, a newcomer will perhaps emerge to buy up for a few cents the shares whose price has plunged because of these difficulties, thereby again demonstrating that financial windfalls are reaped only by those who carefully maintain their distance from work.

A mere nine months after Google Day, Yahoo, soon joined by Microsoft, grabbed center stage by taking advantage of its chief competitor's blunders and setbacks to put together an attractive and reasonable idea. Baptized Open Content Alliance,* this online library project would be free and more timid in all respects, formed by books from the public domain augmented by those that generous-spirited

*This could be described as something like the Free Content Alliance, although it is still too early to say whether it will be the front for future battles against copyright or an entity that aims to profit only from things that cost nothing. The Universities of California and Toronto, the substantial RLG (or Research Library Group), as well as the British National Archives, are all part of this movement, which has extended an invitation to all possible participants to become members (the website address is at the end of the book). Glory is assured to the British Library, which was the first European library to join.

copyright holders might care to offer. Furthermore, the archive could be found on any search engine, something Google had not considered; it was taught a lesson here, as it had never dreamed of offering such openness, frozen as it was in its conviction that all its competition would soon be eliminated. Also, Brewster Kahle, whose Internet Archive was—I could almost say obviously, as we shall see—part of the venture with other powerful partners (HP, Adobe, and so on), extended the hand of welcome to the Mountain View firm and suggested it do the same in a spirit of real enrichment of a great global resource that would serve everyone. As the *New York Times* noted on October 3, 2005, the accord among such high-level businesses, universities, and organizations in pursuit of a nonprofit objective is a rare sight. But isn't the Internet always promising things that have never been done before? After this inviting observation, the question was raised whether the grouping of all these heavyweights and cyberstars, in the aim of merely hatching a project of 150,000 public-domain books, wasn't fairly skimpy, if not downright derisory, in contrast to Google's inordinate scheme—all the more so, as many of the books the new entity's legal advisers had gone over with a fine-tooth comb have had some chance of already being online, thanks to the good efforts of modest but enthusiastic and lively independents.

Other "alternative" enterprises, minuscule or of some consequential size, did not wait to get started and were already forging ahead. For example, Carnegie Mellon University, in Pittsburgh, was soon taking over on a large scale in India and China, with no concern for what the Swiss or German robots cost. Here the pages were turned by hand for an average salary of some forty cents an hour. With a hundred employees already hard at work in India at the end of 2003, the Million Book Project felt it was living up to its ambitious name. Additionally, as noted by the academics managing this undertaking, the texts of India were written in fifteen hundred languages and seventeen alphabets. In comparison, the classic Chinese literature, also being digitized in this same project, proved to be of a charming simplicity.

But this news had barely emerged when the National Library of China announced that it had just asked Google to put into bytes for free distribution 25,000 stampings,* 100,000 texts from Dunhuang, 5,000 manuscripts from the Xi Xia Dynasty (1038–1227), 6,000,000 periodicals from the Republic, and finally 330,000 classic Western printed books dating from 1473 to 1926 "that the library has purchased" (a curious qualification). This communiqué from the December 21, 2005, *Shanghai Daily* raises more questions than the certitudes it offers; but while awaiting further clarification, it can be noted that, in addition to the intoxication furnished by one-upmanship in the land of propaganda, the times of great leaps forward cause those ruled by vacillation to be quickly forgotten.

*Paper rubbings of ancient texts carved on stelae; quite often an ancient stamping will survive when the stone it copied has vanished. The fabulous value of these objects stems from the fact that the sculptor scrupulously reproduced the calligraphic and personal form of the brushstroke.

11

Concordant and Discordant Clues

Google rushed a certain Adam Smith to the Twelfth National Congress of the ACRL, the Association of College and Research Libraries, held in Minneapolis on April 10, 2005.

Beyond the symbolic name of this spokesperson—the namesake of an eighteenth-century Scotsman who was so bored in Toulouse he began to write the text that would become the foundation of absolute laissez-faire economics[1]—his resemblance to a young Hugh Grant was noted by more than one of the women present. In his admirably transparent remarks concerning collections of books in languages other than English, he observed: "We cannot solve all the world's problems, but we are aware that they exist and we are working on the matter."

Two months later, at Troyes in Champagne, two steps from Clairvaux, its manuscripts now replaced by convicts, the firm's sympathetic representative was John L. Needham. With steely gray-blue eyes and the kind of wooden smile characteristic of those executioners for a good cause sometimes invented by the cinema, he presented himself as a book person and spoke about his children. His evangelical attitude was flagrant, my neighbor noted, if only through the otherworldly nature of his remarks. His talk was a long credo in which every sentence began with "For us": "Our engineers have as much influence as our managers . . . We recruit the best students . . . Our users are our first priority . . . Pertinence and neutrality . . . Our values: cultural diversity, exhaustive scope, concentration on the user's interest . . . In short, for us, the book is more important than a Web site." Nothing in all this is truly false, which is where the real problem lies.

The missionary mentality has been imposed on the firm starting with St. Sergey Brin, its boss—who, fortuitously visiting students at Berkeley in the autumn of 2005, a thirty-year-old wearing (if one can call it wearing) an old T-shirt and prefaded jeans—professed that it is possible to "make money without doing evil."* The main thing is to believe in what you are doing and not worry about the competition. When he left after an hour of relaxed dialogue to go home to his ten billion dollars, he left behind a campus won over to his point of view.

The essential point of his message: Google Print is giving the world "a catalyst that does not compete with ongoing digitization efforts" and even offers European archives several different forms of aid to choose from: free dematerialization (the replacing of "material" books with their electronic equivalent) but also shared expertise, accommodation, distribution, and so forth. In short, this was an attempt at a point-by-point response to all the anxieties the old folk had been experiencing since January. The planet recovers its serenity—but it turns a deaf ear.

While the operators currently holding the upper hand have an easy smile but guarded speech on what is really happening, either as strategy or because they genuinely do not know, information concerning the great digitization is suggested to us indirectly—one could

*The new economy, in fact, boasts of being inoffensive in comparison to old-fashioned polluting, exploitive, murderous industry. But Yahoo, single-mindedly pushing its big plan to penetrate the Chinese market, had no hesitation about turning over to the Chinese authorities a journalist who mentioned on his blog police preparations against any potential Tiananmen commemorations. This man was given a ten-year prison sentence. When questioned about this, Sergey Brin responded that his company obeyed all the laws of its host countries. Self-censorship contractually guaranteed.

almost say by the band,* giving this new economic order a whiff of the Cosa Nostra.

Everything that has been free for the taking since the time of, say, Noah has been given a price tag in the space of one day to the next. This includes such things as water, air, even the mere act of looking, surtaxes on telephone calls, and the administration of providers of recondite services that are already too expensive. This is known as playing to win: by definition, any boycott movement cannot originate in the exploitation of sheep by other sheep. So how can an activity as weakly defended as public reading escape this widespread and vicious takeover?

Here is one vaguely traumatizing example: It was suddenly discovered that a number of French texts have already been digitized in various English and American universities. The University of Virginia offers works in eleven languages, among them Apache, and in XML,† if you please. So here we have a mouthwatering "electronic Voltaire" that has taken a detour from its catalog. It amounts to 1,079 shelf marks and documents, but, just like the other desirable treasures in this archive, it is accessible and free only to its campus subscribers: the university bought it[2] in Great Britain‡ for their exclusive benefit. So while the Gauls are sleeping in their old villages on the idea that Voltaire is their personal property as well as a legacy they generously offered for the world's consciousness, the profit-makers who know no boundaries are quietly appropriating it to transform it into a commercial product that anyone with extensive means is authorized to acquire, which essentially amounts to forbidding the majority access to it. Of course, you can still

*[An untranslatable pun; the word *bande* in French means "band," as in bandwidth, as well as "gang." —*Trans.*]

†[Extensible Markup Language. —*Ed.*]

‡On the same shelf, we find a futuristic "Shakespeare Collection" by competitor Thomson Gale, presented as the last word in intersecting research. There is a demonstration at www.gale.com/Shakespeare/tour.htm of this authentic labor of abyssal indexing, which transforms the Shakespearean galaxy into an object of contemplation, study, and reflection akin to an armillary sphere. The absolute has a price: three thousand euros a year.

buy Voltaire's books in paper—although the selection of his other titles is shrinking in proportion, it seems, to the increasing number of different editions of *Candide*—or else head to a large library, while it is still open, to read them for free. But our own venerated National Library, as we have seen, has already begun to get its hands dirty.

Another sign? His worshippers depict Bill Gates as a saint. Oh, yes, he is clearly the king of charity that begins at home: since the beginning of the nineties, his Corbis Society has taken full possession of images that were until then accessible for free, or has bought them for an obscenely low price from iconographic collections of universal interest—such as those of the Hermitage and the National Gallery, among many others—to create, following digitization, a gold mine of royalties worldwide for his sole profit until the end of time. Sell things that were available for free—that is the man's true genius. One question arises: Why did it take thousands of years before anyone dared do this?

One final and revealing example. The Amazon experience speaks volumes. Going beyond the sale of books piecemeal by the page or of e-books, this mail-order bookstore is gradually heading toward printing out-of-print books. All it requires is lending an ambitious ear and adding to its list the inordinate number of books slumbering like *Sleeping Beauty* on library shelves, forgotten sometimes by even their authors, to see profits climb from the million- to the trillion-dollar level.

"Why do we need publishers?" Amazon seems to be thinking.

"Why do we need Amazon?" thinks Google.

12

The Big Picture

Lise Bissonette, president and general director of the National Library of Quebec, which opened on April 29, 2005, is worried: "Today, Google imposes its advertising upon us. But nothing says that one day it won't demand an entrance fee. Now, the essence of a heritage library is that it is free of charge." The opposing view is expressed by a young librarian of Seine-Saint-Denis (Luc Bellier, in an interview). It is not impossible that this essence is the key property of the search engine, at least as long as that search engine enjoys a monopoly on Internet access. Because it is free of charge, it is able to offer anything and everything, which then allows it to sell advertising more effectively and for more money. The exponential growth of the firm's complementary offers can be regarded as a possible guarantee on this point, to the extent that the digitized legacy is in the midst of becoming, for the public and the professionals, a completely separate kind of media, one that is "an entirely new type."*

"And if Google goes bankrupt, who will then own the digitized heritage?" Montreal insistently asks. A clause in the contract with the five large libraries specifies that they will receive an electronic copy of any book they've digitized. Ptolemy did the same thing, but did not return

*As François Mitterrand said when he disclosed plans for the new French national library, twenty years ago.

the original, according to legend, which is not the case here. On the other hand, it seems likely that the partner libraries will not be authorized to cooperate with Google's competitors nor, perhaps, be allowed to personally distribute content that they provided and which, for the most part, belongs in the public domain. Did they truly grasp this? Michel Fingerhut, director of the multimedia library Ircam (Institut de Recherche et Coordination Acoustique/Musique),* also recalled in this regard the U.S. Treasury injunction to American publishers in March 2004: "Any edition with contents coming from nations under U.S. trade embargo (Sudan, Libya, Iran, Cuba, North Korea) must receive authorization beforehand from the Office of Foreign Assets Control (OFAC), the division charged with supervising the strict application of commercial sanctions." This kind of arrangement, Fingerhut explained, could be used to either forbid a provider from distributing contents from a trade-embargoed country or the opposite, forbid providing these countries content that had been digitized in the United States. The ideal would thus be for Google to be a portal that was remunerated only using a financial device similar to the ancient *octroi* tax,† and for all its content to be stored elsewhere, such as the originating libraries. This may not be such a good idea, as the White House is not above all suspicion, but maybe storage could be spread throughout the entire world, as H. G. Wells suggested. In the present case, the contract should specify that the keeper of the texts would recover distribution rights in the event of bankruptcy or some major contractual violation.

"In Mountain View," says one Susan Wojcicki, "we have a gift for doing things in a big way." Here, at least, is something that cannot be denied, just as it is undeniable that it is currently impossible to measure the weight, the cost, and the maintenance know-how sought by the planetary memory in the process of shaping itself. Up to now, this famous search engine could function, if we understood correctly, with

*[Music research library associated with the Pompidou Center in Paris. —*Trans.*]

†[*Octroi* is a local tax levied on the entry of goods into a municipality or any other specified jurisdiction for use, consumption, or sale. —*Ed.*]

a very large number of small servers considered to be disposable: eleven thousand units in 2002, and ten times that number by the end of 2003, disseminated in "farms," four secret sites in California and Florida. But could the colossal internal capacity demanded by the new project be satisfied with some brilliant tinkering?* The other indisputable fact of the electronic world is its own obsolescence: in ten years, none of the computers today will be compatible with the systems yet to come. As the conservators of modern libraries are fully aware, to guarantee the durability of a text foolishly removed from its paper support means addiction to a Niagara of endless expenses. From the pitiable microform to the floppy disk, from the hard disk to the nanochip, a popular novel costing five cents two hundred years ago will eventually surpass in additional costs the value of an incunabulum, although it will still be readable in its modest original form.

Once the smoke from the backfires had subsided—for the most part around February 2005—the silence of the tomb suddenly fell over the real players in this contest, adorned here and there with festoons of evasive rhetorical language. Everything transpired as if the firm of Page and Brin, whose plans had been dampened because of the uncontrollable if not hateful fallout in the mass media, by the increasingly cruel cost of the operation, and by the legal and technical implications that had perhaps not been fully weighted, had decided to rethink its estimates and objectives. It was no longer a question of simply fine-tuning some "mysterious know-how." Unless the quake had simultaneously given the NASDAQ Exchange, the press, and foreign librarians hysterics. No one noted, though, that perhaps a misdeal was also involved.

The budget of "$150 to $200 million," for example, provided by the digitization plan released on December 14, 2004, represents a tidy sum, certainly, but a ridiculous amount on the European Union scale

*As a noncompressed book requires around 6 GB of memory, a million books would then weigh 6 petabytes (millions of gigabytes, each of which consists of a billion bytes). One company, Capricorn Technologies, ipso facto hatched out to launch production of the "PetaBox." This laughable name has been trademarked.

(and that of many other grandiose examples of waste) to have warranted all the editorializing and opinion-making by individuals who more resembled jack-in-the-boxes than reporters in a press that is often writing its stories under orders and is less and less apt to verify its sources. But it is erroneous. Google never took the risk of providing any figures for anything. These alleged ten-dollars-per-book figures for digitization appear in a *New York Times* article published the day after the official announcement of the project. This is what it says:

> Although Google executives declined to comment on its technology or the cost of the undertaking, others involved estimate the figure at $10 for each of the more than 15 million books and other documents covered in the agreements. Librarians involved predict the project could take at least a decade.[1]

In substance, the heads of Google refused to discuss the technical and financial details—"others involved estimate the figure at $10," and so on. Not one single denial or clarification has ever been made about this start-up, and disinformation continues today to give birth to exponential effects, although the specialists deem at least one hundred dollars, if not one hundred euros per book to be the true price of the operation. Obviously, the fine-tuning at a forced march of a new procedure, justified by the scope of the decision and the authority to give the order, could not help but lower this figure spectacularly, but undoubtedly it will not be 90 percent less. As some of these experts, whose experience and expertise are beyond question, work at or for the BnF itself, quite a few questions can be raised on the ins and outs and aims of what could only be, if not an unconscious act, a manipulation.

But if some sort of dirty trick took place in Europe, it was perhaps supported, consciously or not, by a Californian bluff: in all likelihood, Google was in no position to foresee or supervise anything. It has been obvious, for example, since the beginning of 2005 that the name of the program would lead the public to think that the books could be printed (Google Print*)—which could be the case, perhaps, on demand, as could easily be guessed, but only later—and the organizers needed almost a year to change and subdivide it into Book Search with regard to its offer to publishers and libraries.

The more-or-less anonymous directorate of these operations would therefore have to be satisfied with the effects of the announcement, and it would have to navigate by sight in the technical domain, and blindly in the legal domain, while waiting for mass digitization to become easier and less risky.†

The very thorny question of the financial and intellectual rights of the authors or their representatives lies waiting, in fact, for the chance to trip up the would-be mass digitizers.

A book enters the public domain at different times, depending on the country. It is fifty years following the death of the author, according to the Bern Convention, and somewhat more in Europe and the United States, as will be shown in detail further on. Added to this is the moral copyright: even when a book is too old to warrant any financial remuneration, respect for the integrity of the complete text must be guaranteed for now and always. This indisputable constraint is recognized only in Europe and truly followed to the letter only in France. I am told, however, that the book is governed by the laws of the country where it was first published. But allow me to add one more word to the above: this could mean all French books might be pulled off the international stage.

No fewer than three interfaces are necessary, the missionaries finally confessed, to sort through all the complexity.

The first, and the clearest in principle, is Google Print Publisher. It

*To try to preserve some clarity, I have retained the name Google Print for *The Great Digitization.*

†Despite a stubborn appeal, the European protagonists—J. L. Needham (Google), Ivo Iossiger (4digitalbooks, a machine manufacturer in Switzerland), and Ronald Milner (the Bodleian Library at Oxford)—sent back a message with the unanimity of cement for the whole year of 2005: they had nothing to show. So one is free to imagine whatever one likes.

is also the easiest to be realized: if Amazon can do it, Google can too. Their starting principles are also identical. The arrangement they have offered publishers gives free advertising to the book, to which has been added the fairly vile carrot of a share in the advertising receipts from the screen margins, although the presence of ads can be refused as a contractual condition. The reader, for his part, asking his fine mirror, mirror on the desktop about the presence of a word in a given book, for example, "stakes," within the work of whatever peacock du jour* he chooses. This triggers the appearance of a long list of its occurrences, and all that is required is to calmly read each page concerned, as well as the two pages before it and the two pages after it, for a total that, in principle, is limited to 20 percent of the book (although the figure can vary depending on the contract). Following which, the diehard Internaut can finally set in motion the online purchase procedure, or see whether his favorite bookstore, perhaps right on the corner of his block, might not have a copy available in an hour's time. This assumes that the very small booksellers will find out about it and plug in, but why wouldn't they? It is almost a sure bet that this would not cost them a penny.

There are then two cases that come up for this system as it will be applied to libraries.

The first proposition provides access to the entire book on a page-by-page basis. These are books whose copyright has lapsed and which are now in the public domain (publication before 1923 in the United States and 1900 elsewhere, Google's operator has announced incorrectly—he should not be long in realizing his glaring error). Dozens of sites are already offering this; they will no doubt soon number in the thousands.

Finally, a special solution is intended for books still under copyright

*[*Faisan* (pheasant) is a French term used to describe a showoff, cultural crook, or philosophical careerist, the salient trait being someone who talks too much. "Peacock" more readily conjures up that image for English speakers, with "du jour" adding the notion that the fame of such an individual is fleeting. —*Trans.*]

but that have become unavailable and of which, sometimes, only libraries have a copy. The search engine is gradually realizing that problems lurk here. Not having any desire whatsoever to seek an authorization that is impossible to obtain unless one turns the graveyards upside down, the search engine is satisfied simply to offer a bibliographical note accompanied by a three-line extract including the key word the reader entered. This is the *snippet,* a short extract from a book, the term being derived from the verb *to snip,* meaning to cut with scissors. In reality, it is like a strip carelessly torn off a page that gives an image of something so ugly, loose-fitting, and annoying that it is impossible to believe it could inspire anyone to read more of the actual book from which it was extracted. While accepting that this is the case, or while expecting that a remedy may soon be devised to fix this visual disaster, what are we to do in the meantime? The closest library is pointed out, of course, although it is sometimes at the antipodes. But then what? Well, then, let's consider in whispers that in the case of out-of-print books, which are expensive to digitize, it is impossible for the operator to avoid the thought of printing them, even just a minuscule number of copies, perhaps even just one, because such a thing is now possible. If the author receives compensation, who is there to take umbrage? Not the publisher; the book was pulled off the market when he ceased to sell it. At the end of the twelve months following the time a book has vanished from the bookstores, an author can recover full rights to this property. No house would dream of leaning on some yellowing contract as a means of sealing off access to a book—who wants to be branded as petty or a censor? It is the bookstore that risks shedding a tear, even if it is two copies: the print run is of no consequence.* On the other hand, if the operation gives new life to the book, it could entail a return to a paper edition with either the original publisher or a new one.

*The trade should not neglect to take an interest in machines that allow instantaneous printing of a perfectly presentable copy of an out-of-print book. Therein perhaps lies an opportunity offering great advantages and an answer to many future questions.

In all three of these hypotheses, it should be possible to index the entire book, whether by Google or its colleagues; otherwise the system will be in vain. It is this total digitization that alarms the paranoids who dread the future without having any clear idea why.

The day Google Print was introduced to French publishers, they came en masse and left shrugging their shoulders. "While the smaller publishers, notably those specializing in the human sciences, economics, law, and so forth, generally displayed how greatly this means of promotion appealed to them, the larger publishers reserved their judgment. They are quite aware that the wealth of their catalogs will supply the base of Google, but wonder what profit if any they will earn from it."[2]

The day Google Print consulted French writers, no more than six authors, along with four legal experts and a journalist, responded to the SGDL's [Société des Gens de Lettres] invitation, although it was accompanied by a nice buffet. They soon had to face the harsh truth: their contracts forbade them from taking any real interest in the matter; the only right left to them was the right to have a coronary over a typo during the reproduction of their opus.

In the first case, the authors ought to have been rejoicing because their publishers were suddenly assisted by ample means at the very moment a book ceased to interest them—which could be the very day a book goes on sale for the largest publishers. And in the second case, the fact that books are being fished back out of the libraries? Authors should have been jumping, this time for joy, to the heights of its Charles X–style molded ceilings, but this was far from the case.

In the Hotel de Massa,* aloofness is de rigueur. However, the older writers and even certain authors of forgotten essays should quickly grasp that it is now in their best interest to deal directly with the cyber companies in order to see a rebirth of their work, if they deem it still worth reading and, incidentally, can earn some money from it. Similarly, the societies that protect writers should not fail, after the due diligence

*[Headquarters of the literary society, the SGDL. —*Trans.*]

expected of them, to dive into the breach to breathe new life into the works of those who have vanished—inasmuch as that is their job.

So let's leave the manufacturers of the condensed edition to play with their armies of trucks and inundate the rail stations with their spoiled paper. Let's keep digging our way to more luminous outlets. In brief, it is only for the moment that writers by trade and a few independent publishers are in a position to rescue this poor Google Incorporated that has gotten so bogged down in a project that is well beyond its current capabilities but is one the future truly needs.

13

First Trials

The four musketeers affiliated with Google Print Library are, as we know, five in number. The last is the only one that is not located in the United States; it is in Oxford, England. Among the first four archives concerned, the library of the University of Michigan* deserves particular attention, because it was the only one that offered the whole of its collection to the project—some seven million titles, including books under copyright. It was also the only one of the four to be a public institution, in consequence of which it immediately experienced mortal agony from questions on its complicity in a possible violation of section 108 of the American copyright law, which stipulates that libraries have the right to reproduce one copy of a book and give it to the person looking for it, "without any purpose of direct or indirect commercial advantage." Now what could be more indirectly commercial than the page of a book—as was shown at the end of 2004 by a very proud and confident librarian from the University of Michigan Library, John Wilkin—accompanied by advertisements and "sponsored links"?

Peter Givler is the director of AAUP, an association made up of twenty-five American university presses, which are nonprofit companies.

*The word *Michigan,* I've been told by an old librarian, means "cracked" in Yiddish. [(Per Wikipedia) Michigan "was named after Lake Michigan, whose name is a French adaptation of the Ojibwe term *mishigami*, meaning 'large water' or 'large lake.'" —*Ed.*]

On May 20, 2005, he wrote a six-page letter to Mountain View and raised sixteen questions concerning copyright and book reproduction, questions he said he hoped would be answered in June.[1] The following August 11, he had still received no response, but Google suddenly announced on its official blog that for a ten-week period the library books still under copyright would not be digitized and that the publishers could take advantage of this delay to voice any objections to having this done. This perverse arrangement, which has become the final moral court of modern societies, is called "to opt out." It allows the person who exercises it to act as he pleases with regard to someone else, unless the latter has previously and officially spelled out his disapproval. As for anyone who doesn't let out a peep: consensus!

Was this the beginning of some endless spiral into a legal maelstrom? Google's detractors cried victory, especially those who had trouble reading English. In reality, the celebrity lawyers for the big copyright arguments felt that Google could win this kind of trial, because the judges were likely to rule the social value of the work being placed online was more important than the profit that could otherwise be generated by those books, which remained small or adventitious. One of the lawyers, moreover,[2] demonstrated the absurdity of the claim: if authorization was required for each book from which extracts were to be published, it would be akin to asking permission of every property owner to draw a map of a country. We might add, 100 percent of authors want to be discovered by readers, but once discovered, they want to protect what they have created, down to the last dime. Only 10 percent then make the law for the others (only 2 percent of the twelve million books available for sale in the United States in 2004 sold more than five thousand copies).

The following September 20, Google Print was nonetheless hauled into court by the Authors Guild and three of its members—Herbert Mitgang, born in 1920, former Army journalist, former editor of *Stars and Stripes,* and a former president of the Authors Guild; Betty Miles, whose birthdate was carefully concealed, author of thirty targeted bestsellers, including *The Trouble with Thirteen;* and Daniel Hoffman, born in 1923, an official laurel-wreath-crowned poet (his book *Poe Poe Poe Poe Poe Poe* was a National Book Award selection in 1971). The guild had already attempted to obtain money from the New York Times Company and Amazon.com on the same basis. It swore that it and its eight thousand members had nothing against the inclusion of their works in the projects in question if they were appropriately contracted and offered some dollars as well as some free publicity. This assertion was part of five talking points that the subscribers to its information letter were invited to learn by heart and echo if they were "headed to a dinner party this weekend" (September 23, 2005). But "real writers want their works read," retorted David Youngberg, who went on to say: "Instead, you are betraying the craft by attempting to extinguish the free flow of information." This testimony, obligingly supplied by Adam Smith of Google,[3] takes a position, like all that preceded it, on the side of reality: the three old duffers and their club of imitation authors—for it is not enough to write and sell books to become a real author—did not stem the torrent. To the contrary, they simply cut themselves off from a future that will witness the resurrection by libraries of a great many slumbering books.

Likewise for the five large publishing houses that rode to their rescue and attacked Google on October 19,[4] the day following the release of yet another trimester's financial statement boasting higher earnings. In this battle of sharks versus piranhas, this time the formulation of the difference of opinions placed more emphasis on the interpretation of *fair use,* the convention that allows text to be cited and even duplicated as long as it is not for money. As can be seen, the publishers had complete confidence in Google's genius to make large amounts of money at their expense, and they did not fear sealing off their own activity somewhere where it would remain safe from evolution.*

*At the time this book went to press, Google decided to share a few million dollars with its opponents so that they will calm down and, I suppose, do not miss the train in the future.

Although it smells somewhat like the poorhouse or a dead end, this news from the world of the Muridae will be followed vigilantly. It contains the possibility that literary criticism or university research will one day be forced to pay for merely mentioning the least significant book. No doubt the price charged will be cheaper for those lavishing praise on them than for those seeking to demolish them.

Whether this or that company succeeds, in record time or not, alone or jointly, in putting together a colossal collection of virtual books, the sole certainty is that it will happen. It is not merely because it is desired by economic logic, a taste for sharing, or national will. Already taking a position in opposition to the legal experts is a band of libertarian rebels who are shaking the blogs, advocating that one proceed full speed ahead implementing digitization without limits. Taking umbrage, in sometimes quite sharp terms, these rebels view the sight of the great librarial avatar being obstructed by haggling as incomprehensible and without a future. So a large crowd is already knocking the greedy and petty organizations out of the way and preparing to take the great leap forward.

The next revolution will be cultural.

But while waiting for it to get here, let's inventory the advantages and ravages of the intangible.

14

Burning Stakes

What do French librarians think of this phagocytosis? Not a great deal at the moment, all the more so as others are accustomed to speaking for them.*

As scrutinizer of what is brewing on the net since its beginnings, "information science" is a second cousin to the all too often disparaged education "science." It has moved out of education's shadow today to split hairs at the urgent invitation of a fully fed-up world. Its utterly

*Librarian commentators who are both courageous and informed are rare: Dominique Laharay, Michel Fingerhut . . . The others—like the talented Hervé Le Crosnier, *maître de conferences* [tenured position similar to associate professor —*Trans.*] of "information management" in Caen, Jean-Michel Salaün, Hubert Guillard, and Olivier Ertzchied—are often specialists in areas of expertise that are quite foreign to the library, if not its exact opposite. Torn by a generation gap and societal conflict that worsens the mutation of local public reading, the academic ENSSIB (École nationale supérieure des sciences de l'information et des bibliothèques, the university level for librarial studies) is notoriously locked outside of the most crucial debate it has had to face, as can be verified by examining its official organ, the *Bulletin des bibliothèques de France.* As for the ABF, its librarian members debated the right to read in June 2005, then organized a day of information and reflection for October 10, 2005, on the serious topic of digitization, into which they are now crashing headlong. The user hopes strong positions will be taken, but the non-publication of the most highly awaited proceedings of the congress can be seen as a sign of a strange disarray—as shown, for example, by the attempt to keep its "Information Day" journalist-free by charging reporters one hundred euros to attend, a fee that not only appears brutally steep but is also quite at odds with the meaning of the battle just declared.

downplayed reaction to omnigooglization* is therefore carefully gone over with the kind of fine-tooth comb appropriate for the subtle articulation of ideas it employs and, in imitation of all the specialists of "the economy of attention" who wish to be found credible, the rapprochement of terms smacking of the prototype. An "iteration of informational entropy," for example, will impress provincial students more than the repeated disorder of the ideas besieging us. But sometimes even the most compact discursive jam allows some unintentional clarifications to escape, in which is slid a "geopolitics of knowledge" with "European stakes" for "a movement of the nonaligned powers" of learning. It is easy then to guess that the infoscientist will go arm in arm with the cultural notable to patrol our borders. But perhaps he is only doing this in hopes of an opportunity to rifle the latter's pockets.

Serendipity is the noun coined by the irreplaceable Horace Walpole from a Persian fairy tale translated during the sixteenth century. Its title is "The Journey and Adventures of the Three Princes of Serendip," whose heroes are always extricating themselves from scrapes by using their powers of observation instead of the thorough intellectual education from which they had benefited. Although the word conjures up something like the notion of greatly frustrated pity, the Quebecois have adopted this *sérendipité* (their engineers, however, prefer the amicable *fortuité*) to designate unexpected discoveries and say, quite a lot, "He who is not looking finds." The notion caused one observer of the new resources[1] to observe that it applies equally well to the exploration of the World Wide Web—in particular, one might add, when one has been knocking oneself out for hours trying to absorb recondite 6-point pages and a kind of hypnotic state has finally convinced one that the truth lies right before one's eyes. Then all that remains to be done is to make it fit with the chain of reasoning already in mind, with the hope that it may still be there. Should our electronic maxim instead be "He who

*The term *omnigooglization* was coined by Eric White, of the Bridwell Library in Dallas, at the end of 2004.

looks finds something else"? This sounds just as marvelous and is even a bit reminiscent of dear objective chance. But according to the guardians of the established order like Michael Gorman, the new system would be an idiotic pickax that comes up with nothing but uprooted snippets lacking any context and real value, whereas the traditional card catalog, once computerized, would involve knowledge of the global meaning of a work in its entirety. We would then be embarking barefoot and clad in only our nightshirts for the teeming galaxy of information while renouncing understanding it and dominating it from the start. It is a case of Alzheimer erasing Gutenberg, curses the grouch remaining on the tarmac.

15

Advent Eve

The mysterious and deeply overwhelming nature of what we are living through seems to have excited the leaders more than it has the public in the middle of the 1990s, the opposite perhaps of what is currently taking place. The archives from this still recent past also give off as much charm and as many teachings as the testimonies of the copyists working during the time of Constantine the Great.

Relegated to the furniture repository of chronology between bebop and e-book, here, for example, is the computer-assisted reading station, or PLAO (*poste de lecture assistée par ordinateur*). Serge Solomon, who was assistant to the director of Computer Science and New Technologies at the National Library of France, praised its imminence this way:

> PLAOs are stations dedicated to *lecture savante** of digitized documents assisted by computer. They are set up on the garden level in compartments. There is a fee for the use of the PLAO. The stations can be reserved. The PLAO offers all the services of the PLS [*poste de lecture-simple*] with additional features. The PLAO therefore

*[Scientific or scholarly reading. —*Trans.*]

> permits reading and writing with common functional features of existing software (browsing, marking, structuring, text analysis, text search, and so forth). It also offers optical character-recognition tools. The user can build his own body of work from a variety of sources that are offered: digitized BnF documents, CD-ROMs, and free databanks on the Internet.[1]

With a more than fifteen-year head start, the possibility was thus offered to become acquainted with the interstellar voyage of words that was still called *lecture savante* and not yet hypertext. "It was a realization of the first order, based on a profound, preexisting theoretical reflection, which is not always the case in the domain of documentary computer science. . . . This venture retains a distinctive brilliance," testifies the most up-to-date and complete guide on the digitized library, the Jacquesson-Rivier.[2] But how is it, you might ask, that this country is consequently not a pioneer in this field? It is because we have decided not to proceed beyond the theoretical approach. And why is that? You have to ask the hierarchy of that time, who are all today no doubt enjoying a well-earned retirement. Serge Solomon also formulated the way we could prepare for the time bomb that, to our great surprise, is exploding today:

> How is it possible to extend and provide more widespread access to primary information by taking advantage of technological advancements that will make it possible, within the next ten years, for a billion human beings to benefit from the services of future "information superhighways," while preserving the legitimate interests of producers, authors, and publishers [of creative and/or scholarly works] or of the museums and library charged with administrations of their national legacies?

Yes, this was the Middle Ages, and people spoke with shivers of the "electronic scriptorium." For example, Yannick Maignien, another member of the team, said he smelled out in the stale ideas of Michel Foucault[3]—the herald of the scriptuary compote now available, moldable to every reader's desire, who would then tend to become the author himself, on condition of anonymity—"the disappearance of the author function in a way that will once again permit fiction and its polysemic texts to function anew in another style." In other words, "One can imagine a culture in which discourse circulates and is received without the author function ever putting in an appearance." This allows Yannick Maignien to conclude, "The PLAO and digitization are 'machines' for structuring knowledge, just like the codex, the scrolls of Alexandria, or the alphabetical entries of the Encyclopedias of Diderot and d'Alembert." They also served to catch the attention of Bernard Cerquiglini, who says, à propos again of this similarity between the medieval collective text and the new work the computer permits:

> The screen, especially, can supply, for this sequence of writing, the sequences of a manuscript, or related manuscripts, which, by form or by meaning, by this play of the repeated statement and recurrence that is medieval writing, have an esthetic bond with it. That would be publishing on a grand scale never before realized yet indispensable, and only today's information technology can give us the means, perhaps even the idea. Because the computer through its dialogic and multidimensional screen restores to us the prodigious memory capacity of the medieval reader, it defines its esthetic reception and is the foundation of the pleasure the reader takes in it. The page would then be turned. Electronic writing, through its mobility, reproduces the work in its very variance. Information technology will rediscover, on this side of Modernity, the path of an ancient literature whose trace printing had erased: and this is truly something worth thinking about.[4]

On page 429 of his *Treatise of Documentation,* Paul Otlet had already put this in his own words. "In its new form, the book will be: 1. in constant growth . . . ; 2. redistributing its elements; 3. in cooperation; 4. in analysis synthesis (tables-outlines); 5. in abridged-development; 6. under the authorized supervision of the large associations; 7. international in theory and application."

16

Muta Solitudo

The most noteworthy absence of any display of opinion will be that of the great mutes of our time, otherwise known as intellectuals, an absence that gives the impression they dream of themselves as marble statues. But librarians, all the same? They have given no more thought on the matter. Or would it be more exact to say, no less?

It has been observed that the movers and thinkers of books are deliberately shoved to the sidelines when it is a question of constructing a library as a permanent structure, whether it is a monumental folly or a minimarket.* Are they deliberately removing themselves from the equation at the time we are really making the transition to the virtual? If we can go by what Arundell Esdaile says, though, "Librarians form a race apart: they seek to preserve for posterity, in one form or another, material that was most often intended by its producers to be ephemeral."[1] We should thus be listening only to them at the present hour, rather than to the organs or organisms that have visited this matter as

*Outside of two or three brilliant but powerless consultants, the most notable exception to this rule was Jean Gattégno, who was moreover rapidly banished from the TGB project at the beginning of 1990. For more detail see his quasi-testament, *La Bibliothèque de France à mi-parcours. De la TGB à la BN bis?* [The Library of France at the Halfway Point: From the Very Great Library to the National Library's Duplicate] (Paris: Éditions du Cercle de la Librairie, 1992).

parasites with views that in no way coincide with the best interests of reading, and even less with those of the reader.

Driven back into the bunker of his glossy, finely bound volumes, one curator finally admits: if librarians do not feel it is their concern, it is because they do not feel this episode involves books at any time, only fragments torn from the corpus.

Certainly, reading online will no longer be presented as it once was. But if one deplores it, one should do all in one's power to prevent it, or otherwise take the best possible advantage from it. In no case should anyone concerned with online reading be keeping his offended silence.

We read, however, that the employees of the great libraries should not necessarily be unduly alarmed for their future, even if they show their concern in a worthy display of introspection. There are too many brand-new texts and too much disorder that needs to be filed and classified, even if ineptly. The blasé agent of the overheated documentary will be there to help circulation: "We have returned to the model of the Alexandrian Library and Ptolemy," Lise Bissonette says, with what may well be false candor. "The role of the library is always to bring order to chaos." She is therefore in a hurry to see if a *cataloogle* will appear and how it will present itself, to the extent that since its earliest days, it will be the record of the collections of five huge heteroclite archives crammed with duplicates that will need to be indexed, classified, and coordinated again.

Google "turned back," declaims one author, who views the matter with the eye of a sports fan from the prow of the gondola and the hierarchical listing. Going by his words, the vocation of millions of titles, once they have been digitized with no obstructions, would be to appear together on the screen (page 60 of his bestseller[2]). A chimerical vision. In reality, books plunged higgledy-piggledy into virtuality will less resemble a hundred million butterflies than they will dozing coelacanths. The secret consists of formulating one's questions in an intelligent and truly malicious manner: if a document answering their specifications exists, it will ascend straightaway out of the abyss. It is even permissible to think that there will be more chances for a surprise than in the worthy traditional archive, thanks to the shimmering of the words it contains, which the bots know how to detect. The fans of gudgeons have dubbed this disrespectful practice "fishing from the gut": it has gone on this way for years on the Internet no matter what engine is chosen, and every day arrives on a tide of pages. The justice of the system is that it will give an advantage to those authors whose vocabulary is varied.

The Web also has the charm that transforms any of its users into an apprentice dowser. But this sorcerer is not left to his own devices: every search engine has a page offering a grid of advanced questions that can represent a fastidious puzzle for the beginner but really allows for the browsing process to be fine-tuned and simplified. No system, however, dares use the adjective "intelligent" in the wealth of technical encouragements it bestows for the moment to describe its process without retaining the quotes.

Yes, the positioning of the intangible book on the modern reading stand can appear to some stability-lovers as implying a previously existing scorn of its general meaning. This is the assumption that allowed catalogs to be drawn up that were as meticulous as a troop of infantrymen and were forever the indispensable map to the treasure on the shelves. But in compensation, this upheaval permits the unexpected and refines searching by content. As the former way of thoroughly exploring texts remains facilitated in every manner by widespread critical bibliographies, how could this be seen as anything but a comfortable addition? On the other hand, the spilling of billions of book pages onto the Web cannot fail to cause the sprouting of a multitude of monumental sites dedicated to their reasonable classification. At once working professionals and retired scholars will be itching to vie with one another, providing generous and useful information and summary notes. This act of compilation on its own also bears the name "library." Photius gave us the first great example of this literary Boy Scoutism in 867.

What would the old sage Lao Tse [Lao-Tzu] think of this, he whose civil profession was guardian of the "archives of heaven" and first member of the corporation? In the shadows, Borges has a half-smile on his face.

QUOTE OF THE DAY

Yahoo *as I am, it is well known through all* Houyhnhnmland, *that, by the Instructions and Example of my illustrious Master, I was able in the Compass of Two Years (although I confess with the utmost Difficulty) to remove that infernal Habit of Lying, Shuffling, Deceiving, and Equivocating, so deeply rooted in the very Souls of all my Species; especially the* Europeans.

Jonathan Swift, *Gulliver's Travels,*
"Letter to Cousin Sympson"

17

An All-Horizons Inventory

Let's now rummage through the most current, very fragmentary and confused, although promising, panorama of the digitization of books.

All I am taking into consideration here are the sites offering free reading; the rest are, by definition, excluded from being libraries, in the meaning one hopes that this word will always be understood.*

Dematerialization is on the march. Let one hundred virtual libraries bloom, which, put together, form but one. The heart of the spider's web that is every user's keyboard at the moment he spurs on his or her favorite search engine can, in fact, give the fleeting illusion of an infinite archive. Infinite? If it were possible to evaluate, *grosso modo,* the quantity of titles that could be opened in 2000, any such accounting today is impossible; the collections have been *mise en abime*† as far as the eye can see.

Institutional and private initiatives are sprouting up all over, their

*Netlibrary, Questia, and ebrary are providers of reading for a fee. Presenting themselves as libraries, they constitute a taste of what the future has in store for the distribution of books and differ not at all, in any case, from e-book merchants like Numilog and so forth.

†[*Mise en abime* is a term borrowed from heraldry that refers to the presence inside an image of its own reduced image (somewhat akin to Russian nesting dolls). Literally translated as "placed into the abyss," this term is now used in reference to a formal art technique in which the image contains a series of increasingly smaller versions of itself. —*Trans.*]

efforts overlapping one another and contributing to the ongoing merger; nor are the amateurs to be outdone. The results of these efforts can be displayed in full-text mode obtained by character-recognition systems or—this is not rare—by a manual capture method marked by an admirable self-sacrifice, but also in only image mode for manuscripts or venerable printed volumes, all of which, taken together, are proving to be of the most exciting variety. Light entertainment reading matter sits next to scientific documents of the highest caliber, all carried to the altar of the Net not only because it is the absolute vector, but also, in many cases, by the popularizer's decision. Incidentally, many of the minuscule archives offer a digitization service today that is clearly more advanced than the services offered by the uncontested elephants in this domain; it is often more generous as well.

Here, then, is a list of collections or portals that lead to this treasure, a necessarily provisional selection in which the English language predominates. With regard to the texts of the French—who lag behind their own authority, as we have seen—a critical inventory has been taken by the literary research site Fabula, to which one is well advised to refer: www.fabula.org/vlib/categorie.php?id=72. The university organization that administers it is associated with the oldest Internet directory, founded in 1991 in Switzerland, whose page can also be visited by those whose curiosity so motivates them: http://vlib.org/InformationManagement. Just like the links contained in the endnotes, all these addresses,* which are as daunting as they are hideous, are listed on a special Web page where one can simply click on them without having to spell them out on the keyboard: www.polastron.com/notesLGN.html. Some of them seem interminable, in fact. They are all, in any case, perfect candidates for dropped letters or characters (note that the hyphen for the URLs printed on two lines here are always an integral part of the address).

ABU (pronounced, they insist, as *abou*), Association of Universal Bibliophiles. Admirable and patient volunteers who capture

*Commonly called URLs, or Uniform Resource Locators.

books in their entirety (288 in 2002, the date they suspended their activity); housed by the CNAM. http://abu.cnam.fr

Biblotheca Classica Selecta, a construction of professors of the university of Louvain that offers Greek and Latin works with their French translations. http://bcs.fltr.ucl.ac.be

Bibliotheca Universalis, a partially open door to a list of national libraries, only two of which have plugged their connection online, the BnF and the National Diet of Japan, where the amateur can feast his eyes on magnificent manuscripts. www.kb.nl/Gabriel/bibliotheca/universalis/fr.bibliotheca_universalis_Collections.htm

Bibliothèque électronique de Lisieux [Lisieux Electronic Library] consists of an already substantial collection of short nineteenth-century texts indexed in collaboration with the University of Toronto. An exemplary initiative. www.bmlisieux.com

Bibliothèque numérique Landowski de Boulogne-Billancourt [The Landowski Digitized Library of Boulogne-Billancourt]. More than one thousand books to "borrow" in accordance with the modalities of any lending library. It is free for residents and requires an annual membership fee from others. www.mobipocket.com/ebookbase/library/landowski

Bibliothèque numérique de la MOM, House of the Orient and the Mediterranean. Around fifty rare and precious volumes in image mode. www.mom.fr/bibliotheque/bibnum

Bibliothèque philosophique offers books ranging from the classic texts of antiquity to the (for the moment) twentieth century, in their original language or in French or English translation. This is a product of the Academy of Nice. www.ac-nice.fr/philo/texts/biblio.htm

Bibliothèques virtuelles humanistes, sixty-three books that have been handled perfectly—soon to be two hundred.* www.bvh.univ-tours.fr

*[As of mid-2007, there are 219 works on the site. —*Trans.*]

Casalini Libri digital division offers publications of the French School of Rome. http://digital.casalini.it

CEFAEL, collections of the French School of Athens. More than 250,000 pages online. http://cefael.efa.gr/site.php

Children's Library has a collection of 2,427 books, including those of the ICDL (see below). This collection of children's books has been digitized by the University of Florida and the National Yiddish Book Center, among others. www.archive.org/details/oacl

CNUM, Conservatoire numerique des arts et métiers, offers free access to the digitized legacy of the CNAM Library and that of the Technical History Center. http://cnum.cnam.fr

Digital Collections Program gathers together all the digitization projects of McGill University, in Montreal. www.mcgill.ca/dcp/projects

Ebooks gratuits is the pioneer in the offering of copyright-free books (but warns users that the laws are different depending on the nation where the files are downloaded, and that respect for these laws is the responsibility of each reader) and boasted in October 2004 that it had received 17,635 visitors and 24,274 visits for 65,076 books downloaded. http://ebookgratuits.com/index.php

Elib groups together the substantial and learned electronic efforts of the Universities of Cambridge, Oxford, Manchester, and Leeds. www.ukoln.ac.uk/services/elib/projects

Etext, the Electronic Text Center of the University of Virginia Library. Since 1992, it has made available seventy thousand complete books in SGML or XML format. During a twenty-one-month period from 2000 to 2002, 8.5 million copies were downloaded by readers from more than one hundred countries. http://etext.lib.Virginia.edu

The European Library is a portal that was opened in March 2005 with planned access then to forty-three European collections, all with their own online resources. More details about this particular subject are provided further on www.theeuropeanlibrary.org/portal/index.htm

Gallica, name of the National Library of France [BnF] collection, described in greater detail elsewhere in this book. http://gallica.bnf.fr

The International Children's Digital Library, founded by the University of Maryland in 2002, has a catalog of 611 books in various languages (two of which are in French, offered by the Swiss National Library), intended for children from three to thirteen years old, which they can navigate themselves. [This site now boasts 2,198 books in forty languages; the two books from Switzerland have now been joined by another twelve titles from France, one from Africa, and one from Brazil. —*Trans.*] www.icdlbooks.org

The Internet Archive is a nonprofit organization in San Francisco seeking to combine the collections of large libraries other than the "Google Five" (Library of Congress, University of Toronto, Carnegie Mellon University, Zhejiang University in China, the European Archive in the Netherlands) in order to offer some million works that are in the process of being digitized, with twenty-seven thousand already available to be consulted. www.archive.org

The Internet Public Library is an immense repository fed by the University of Michigan's School of Information, with a wealth of information for redirecting searchers to the majority of online resources. www.ipl.org

IntraText: Eulogos put together a fabulous collection in Rome in 1994 in HLT and hypertext that is now six thousand titles (50 percent of which, however, are religious) in thirty-seven languages (10 percent in French and about 14 percent in English). www.intratext.com

The Making of America possesses 11,859 works on the social history of the United States. www.hto.umich.edu/m/moagrp/

Medic@, collections of the intrauniversity library of medicine: 3,085 books as of March 2005, the bulk of which are medical treatises from antiquity, sixteenth-century anatomy works, and so forth. www.bium.un-paris5.fr/histmed/medica.htm

The Million Book Project contains 10,611 books. Founded by Carnegie Mellon Institute in partnership with the Internet Archive (Brewster Kahle), this entity offers the distinction of working in collaboration with India and China, which are digitizing their own collections. www.archive.org/details/millionbooks

The Online Book Page catalog and index contains more than twenty thousand books available on the Web. http://digital.library.upenn.edu/books/

The Open Library, born in autumn 2005 on the foundations of the Open Content Alliance (www.opencontentalliance.org), is, in theory, the substratum of the immense reality of tomorrow. www.openlibrary.org

PôLib, the virtual patrimonial library of Lille University, offers an assortment of works from various disciplines, including rare books on Chinese history. http://polib.poleuniv-lille-npdc.fr/index.html

Project Gutenberg is, to some extent, the veteran of this venture, which has enabled it to offer more than fifteen thousand titles in forty-two languages, uploaded letter by letter by volunteers. www.gutenberg.org

Since 1985, the *Thesaurus Linguae Graecae Digital Library* has put online the texts of 3,700 authors writing in Greek from the time of Homer to 1453. Only a small part of this collection is available to be consulted for free. www.tlg.uci.edu

The *University of Chicago Library* offers, through the Digital Activities tab on its Web page, passage to a veritable mine full of collections available in full text or in images, and in many languages that open onto other archives, and so on. A tool of dizzying power, as they are all slated to become. www.lib.uchicago.edu

The World Digital Library (temporary name) is an initiative launched by the Library of Congress at the end of 2005 using donations from the private sector (the first three million dollars was contributed by Google Inc.) to establish a universal world library. www.loc.gov/today/pr/2005/05-250.html

The WWW Virtual Library: Digital Libraries is the fruit of a considerable pioneering effort that was regrettably interrupted in February 2000 (a new administrator will apparently be welcomed soon), but most of whose still functioning redirections lead to no less than 154 archives throughout the world specializing in the human and social sciences. www.indiana.edu/~vlib/Digital_Libraries/

18

The Future at the Portal

Does Europe still have any real role to play in this immense and splendid cauldron?

The large though intangible European library or TEL, The European Library, as well as its name in each of the nine languages of its members (which in French is Bibliothèque européene), is an avatar of Gabriel, a project born during the previous century from a desire for cooperation evinced by all the nations forming the European Union. It opened on March 17, 2005, somewhat discreetly, in consequence of which a number of politicians continue to solemnly pin their hopes on it.

Constructing a continental library, if this restriction makes any sense, means grouping together in the same portal access to the catalogs of existing collections, then to their collections with no delay, without which it would amount to little more than a directory.

> "The European Library Collections" displays the coats of arms of: The British Library, Biblioteca Nacional (Portugal), BN-Opale plus (France), Online-Katalog Der Deutschen Bibliothek (Germany), Helsingin Ylioposton Kirjasto (Finland), SBN OPAC (Italy), Koninklijke Bibliotheek (Netherlands), OPAC Helveticat (Switzerland), and the Slovenian National Bibliography. Russia and its two national libraries should also sign on. It has been noted that

> France joined this group dead last, during the month of June 2004. Yet to enter, according to the documentation, are the base members. These would be the thirty-four other national libraries whose collections would be integrated into this one during a later phase. The daily management (administration, promotion, maintenance, development, conception, editorial work, technical assistance, and so forth) is taken care of by the staff of the Office of the European Library, based in the Netherlands National Library.

The no-fuss appearance of this portal conceals a serious, costly, and difficult labor: coding all the written language systems of the European community in a way that will make it possible to open all the different catalogs crosswise for researchers as well as for the merely curious, and then make available those books in these collections that have already been digitized. It goes without saying that the enormous amount of work accomplished thus far is but the mere beginning of what still remains to be done.

But so much work done under the leadership of nationalism on the part of certain countries and not others seems destined to discredit itself, for what is the reason not *all* the libraries were invited to contribute the results of their own indexing labors to TEL cornucopia? By virtue of such ostracism, will it remain closed to the virtual books of the Milli-Kütüp-Hané of Ankara, Turkey, for example? How could the alleged evangelists of knowledge dare defend an attitude that says, "You are not yet a member of our political club, hence your books are of no interest to anyone"? The most normal librarial gesture, therefore—what one would most expect from the aeropagus of the high librarians—would be to invite with no further delay or beating around the bush all the national libraries electronically ready to enter the game. As a secondary consideration, it would also serve as an opportunity to invent a new name for themselves, one that was more ample and more ambitious.

A visitor recently arrived from the moon would have trouble grasping all that is currently being plotted. But he could at least guess that the governing bodies of France and Great Britain are working desperately to claim all the credit for something that Germany, already offering a representation in miniature of colossal European impotence with its Länder, is no longer in any shape to contest. Thus the French of another age that we put in charge hardly surprised us in 2005 when they went suddenly frolicking like lambs over a fantasy of a digitized European library, which would be, if I understand this correctly, a competitor with the one that already exists and which would be starting from zero without a penny. Baptized BNE (Bibliothèque numérique européenne) and suddenly thrust into the role of the sole political safeguard against alleged Anglo-Yankee intellectual hegemony, this entity would nonetheless find itself generously offered the chance to be headed by the BnF, whose exchange policy is well known. From its four ivory towers over the last several years, it has condescended to allow the establishment of a network known as "associate poles"—a better description might be vassals—with seventy-six "delocalized" archives: municipal, university, specialist libraries, and so forth, which are granted the authority to lend books to the parent library, while it keeps its own collection to itself.

But the 100 percent French Digitized European Library also seems intent on undermining itself with the composition of its "pilot committee" (this vocabulary begins to seem better suited for a submarine . . .), which rolled out the welcome mat for a mob of forty employees from five Parisian ministries as well as representatives from the big bosses of inertia and industry: Syndicat national du livre, Thomson, Éditis, and so on. Based on what the newspapers reported,[1] the presiding minister, seconded by the president of the BnF as vice president, charged the committee with the task of creating "specialized groups" whose mission would be to return to him, within a time span one imagines to be reasonable, with the "necessary arbitrations." Confused by the sight of such scope and courage, the pilot committee envisioned nothing less than inviting neighboring countries to follow its example. It must be admitted that no effort is too much when it comes to drowning a fish as taxing as this.

As a foretaste of what an administration that lets itself go in the cybernetic field knows how to create, there is an interministerial platform created for discovering what riches are to be found at home: *Numérisation du patrimoine culturel* [Digitization of the Cultural Heritage]. Here, "discovering" simply means learning where what you want to find is located: it is still your responsibility to catch a plane there. The Library of Toulouse has digitized four collections, including zero books, but oddly enough they can only be consulted on-site. In this electronic shopping cart, gilded by the social security aesthetic, there were already 132 libraries present, at least in name, as of June 2005. Leaving out Gallica, the treasures of illuminated manuscripts are what shine most intensely at present. Nothing but images, nothing but images . . . In addition to the illustrations of the medieval manuscripts, all that can be found there are photographs and engravings, maps, postcards, blueprints, musical scores, and newspaper front pages. "Digitized" in France, just like "cultural," would therefore have become synonymous with iconolatry, as if, like a father who does not want to see his children grow up, the financial backers do not want to know anything but multimedia. With respect to the printed scientific, philosophical, or literary cultural heritage in the form of printed works being made accessible this way, everything or almost everything remains to be done.

While the BnF could find only a little over twelve thousand euros* in the cash drawer to pay for digitizing books after having paid its watchmen, conversely the Michael Portal (which does not seem to find spelling mistakes troubling), an offshoot of the Minerva project, has thirty-three million euros at its disposal for three years, simply to inventory the growing resources of Europe's virtual cultural legacy. At a meeting that took place on June 21, 2005, in Luxemburg, Viviane

*This is based on its own admission, as we have seen. Fortunately, since the time of that interview, the figure has climbed to 2.5 million. Fog has always had the effect of enlarging the realities at work there. The ministry, for its part, set aside in an envelope marked "BNE" some 400,000 euros to cover 2006 expenses, most likely intended to partially cover the cost of preparatory meetings.

Reding, a member of the European commission, stated that the libraries of the twenty-five member nations held two billion books, 100 million of which were in the national libraries, of which "less than 2 percent" were currently digitized. The proportion is not truly incorrect, but she would have been closer to the truth if she had said "fewer than two out of ten thousand"—this document definitely drowns everyone in numbers—although the cost of the treatment mentioned this day seems correct: from fifty cents to one euro per page (amounting to two to four billion euros to match America's well-known ambitions in this domain). This figure includes the cost of cataloging and maintenance, or at least this is to be hoped. But the essential point of this report was that if the European Parliament and Council had given their backing to lay out sixty million euros from that time until 2008 for multilingual and "interoperable" access to digitized cultural content (program *e*Content+) and another thirty-six million for research (especially with an eye to preservation), there would not have been a single penny for the digitization itself.[2] This would be the responsibility of every member aware of the value of his cultural heritage. In other words, if there was not one or a dozen Googles to make a gift to the planet of the titanic dematerialization efforts required, it would advance only at the snail's pace we are currently experiencing. Viviane Reding reminded her audience of this incidentally but firmly: in Seville, it had already been more than ten years since the Archivo General de Indias had digitized eight million manuscripts and documents with donations from Corte Inglès and IBM Spain, while Telecom Italia was financing the transition of the Sormani Collection into the virtual domain for the city of Milan.

Synergy among different bodies of knowledge is today infinitely easier to achieve than was ever dreamed in even the most delirious Utopias and science-fiction novels. The creation of a link takes all of ten seconds. The ideal situation would be one not weighed down by any hierarchy, protocol, or ulterior political or mercantile motives. Transforming all the resources in the world into communicating vessels constitutes the sole reasonable path toward a Very Great Library, one whose effects we might dream would be shaped by the meticulous impetus of its professional users and the enthusiasm of the curious, like a metastasis of intelligence in the sticky tissue of the Web. The characteristic and often careerist ideological shoulders of a few regional personalities serve more to slow down this evolution than truly to take advantage of it.

This is why, while the blind anthill noiselessly manufactured by investment projects, utopian schemes, and outright rackets continues to swarm and grow higher, the quantity of all the Internet has to offer our bedazzled eyes consists of a still infinitesimal number of books waiting to be discovered and browsed through.

The "Europeana" finally opened its portal if not its doors nor books on November 20, 2008. It offered what the guests had brought: a link to 1 percent of what the continent's national libraries contain, half of which was from Paris.

Barring a miracle, it is a strong bet that the most "endless" of libraries will remain for quite some time to come the one that is made of paper. Does that put your mind at ease?

QUOTE OF THE DAY

When you've seen one pixel, you've seen them all.

Anne Fadiman,
Confessions d'une lectrice ordinaire
[Confessions of an Average Reader]

19

Tomorrow's Readers

It is a waste of time to think the key to this disconcerting gestation is to be found in the transition of the manuscript edition to the printed book, even if in their sluggishness and scope the two phenomena do bear some resemblance. All things considered, the printer only creates the metallic, then mechanical, aping of the *graphein.** Openly snubbed by the great book collectors since the very time printing began to blossom, it has done little but to gradually ruin the livelihoods of the manufacturers of quills and reeds without changing the way we speak one iota. Whether transcribed by hand or by cast-off blocks, the mental springs that moved Pico della Mirandola are as well oiled today as they were during his lifetime.

The revolution most akin to the contemporary upheaval took place much earlier and lasted for two centuries. This revolution involved the formidable transfer of texts inscribed on scrolls—*volumen* or *rotulus,* depending on whether the eye scans the text inscribed on it laterally or from top to bottom—to the codex, which occurred around AD 300. The scroll made from papyrus leaves glued edge to edge was potentially

*Greek for "writing."

synonymous with the infinite continuity of reading, whereas the codex prompted the cutting up of the text into pages, accompanied on the one hand by every imaginable loss and jiggery-pokery—it is as if the whole of antiquity was transmitted through a kind of funnel—and on the other by a new rhythm in the assimilation of the written word, like a brief stroboscopy at the end of every page. While the attentive mind never loses sight of the unfurling text, it requires a sudden turn to keep following it.

So what is the computer doing? Both, simultaneously. It offers the procession, which reconnects with the elevator going up and down of the ancestral *rotulus,* and simultaneously its possible opposite, being able to section things more minutely than the codex. With respect not only to the shape and the size of the screen—it is, incidentally, high time that book "addicts" were offered texts in "portrait" rather than "landscape," to use the inane terminology of the manufacturers—but also to the attention capacities of the person reading, we can make the transition from texts cut into pages to those divided into paragraphs, with any possible foot- or endnotes exiled into a document's limbo but still easily accessible on well-constructed sites through the constant appearance and disappearance of clickable links. Furthermore, each word could send the reader to its definition, or to many other kinds of references waiting in the wings, if not even to other books entirely. The links, "so strong and so tender" as the poet says,[1] will authorize every manner of abbreviation and abridgment. We can imagine, for example, this captivating dialogue:

> "What have you to say about this?"
> "I am only responsible for that."

(Clicking on the first demonstrative pronoun will cause the appearance of a terribly compromising photo, while the second opens an explicatory note that until now has been known as a "footnote," but in this instance may be as long as two hundred sentences.)

Every word of every text potentially becoming the platforms of other realities—nothing, in fact, obliges them to be marked by an underline or by colors—it is easy to imagine that an extremely long online novel would reveal itself only by fragments, showing only its title or its incipit, or even the first word or words: "For a long time," for example, you could also, if like Penelope you have nothing better to do that evening, start interweaving the threads and phrases of several famous novels (but keep your eyes out for the guardians of moral copyright!). Let another generation come to maturity, and we shall truly see in a few short years the emergence of a true cyber-writing, with condensed syntax and frequent carriage returns—but it will be a generation the best of which will know how the new possibilities for increasing meanings can be used to best advantage, thereby authorizing all kinds of spillovers, the mergers of all genres, the intimist narrative perfectly interwoven into the encyclopedic saga and "several other worlds at a time.[2] The triumph of the echo over the report.*

But for the moment, and as a secondary consideration, browsing through the paragraphs is presented by some distributors as an activity flanked by hedges of advertisements leaping like the subjects of a king passing before them in a coach, unless obviously one has found a free provider or is paying for access. All kinds of propositions already abound, as well as doubloons and duplicates in myriad quantity. No doubt we are soon going to be seeing self-publishing microsites. We will then be asked to consent to pay a small fee for the author, who perhaps is lurking there behind his text, in the middle of developing it to its final—and lengthy—conclusion. If this new Dumas loves his heroes as much as his subscribers love them and he loves his subscribers,

*In consideration of this potential, the researchers on the cutting edge even feel that the digitized library is ready to surpass the functions of arrangement and those putting complete texts at the user's disposal to also become a crucible for virtual encounters and exchanges of ideas in gestation: the archive, nonstop genesis. This perspective suddenly transports us far beyond the shortsighted debates of 2005–2006; see, for example, the scholarly article "Qu'est-ce qu'une bibliothèque numérique, au juste?" [What is a digitized library, really?] http://artist.insist.frarticle.php3?idarticle=245#bib8

he will be in hardly any hurry to start envisioning the epilogue.

Those who maintain that the charms of the old-fashioned book made of paper, cardboard, and sometimes leather are unfading, numerous, and widespread, if not still in the majority. People will never tire of this perpetual gesture of the hand caressing the page or, conversely, slapping it when negotiating (as one negotiates a dangerous curve) the rhythmic rupture of the discourse. But those who would cling forever to these delights resemble those portraits by de Chirico, eyes lowered before the perspectives surrounding them, refusing to imagine that the repeated segmentation of the text may one fine day be viewed as obstructive and archaic in comparison with the new fluidity.

So what will it be, *rotulus,* or hopping about like a flea? While the ghosts of Queneau and Roussel are looking over their shoulders, our mutating successors have the choice between the straight line of the sequence that climbs and the zigzag in the transtext. In both cases, pure pixel literature will have a rare word as starting capital: ectasis, or dilatation, this time as far as the eye can see. We could also say ecstasies.

Would we like to know whether it will be possible one day to read Proust and his descendants on the screen? Yes, under these conditions.

20

Last Books! Last Books! Closing Time!

Andrew Carnegie was what the English call a particularly *decent* individual. Born to moderate wealth in Scotland before going on to become the Steel King in Pittsburgh, he nonetheless professed that any man who dies a millionaire was dishonored, a sentiment that led him to disburse $350 million during his lifetime, principally for the founding of public libraries: 2,509 in total, 1,679 in his adopted land. There is one in France—in Reims. The building we owe to the architect Max Sainsaulieu. Inaugurated in 1928, this charming example of Art Deco recently reopened its doors after a renovation that brought it up to modern standards.

A century later, however, we are witnessing more and more libraries stricken by an illness with no apparent remedy, one that causes them to close up shop.

The precursory signs appeared in the United States, when a drop in stock earnings had a direct effect on the monies disbursed to foundations: the overall budget cost of American public libraries was reduced by $111.2 million in eighteen months (from 2004 through the spring of 2005), which brought about reduced hours and outright closings, as well as a drop in the number of books acquired. This sometimes came to the attention of the press but not always, and it required a lot of ballyhoo in the media to thwart the plans of Governor Jeb Bush, brother of then president George W. Bush, to eliminate the Library of Florida and donate its 350,000 volumes to a private university, but nothing was able to prevent the closing (among others) of the Municipal Library of Bedford, Texas, as a budget cut.

> The number of people living in Salinas, California, grew from 100,000 to 150,000 with no proportionate growth in revenues. The result: the three libraries were scheduled to close on June 1, 2005,* unless seven million dollars fell from heaven. Salinas has 38,000 homes in which at least half of the people are under eighteen years old, 64 percent of the populace is Hispanic, and 20 percent of families are below the poverty level. The closing of a library means:
>
> - the loss of investments supported by the taxpayers over the years
> - the transfer of their personnel to other jobs and the loss of their know-how
> - the mandatory travel for the stubborn reader to a neighboring city, if there is one
> - the elimination of a chance for a better education for the citizens[1]

This was how Michael McGrorty reported in his online journal on one of the affairs that left a deep mark on American public opinion in 2004. It should be noted that John Steinbeck was born in this city and that his book *The Grapes of Wrath* had the honor of being burned there in 1939. The emotionally charged citizens of Salinas, including even the inmates of the local prison, eventually took matters into their own hands and raised, a few dollars at a time, the half-million dollars

*[As of April 2007, as reported in *Library Journal,* the Salinas City Council was able to find money in its budget to keep its libraries running. However, they are not out of the woods yet, as they are reportedly having difficulty attracting skilled personnel to staff the libraries. —*Trans.*]

they needed to keep the libraries open twenty-six hours a week until the end of 2005. Then . . .*

In addition to this fairly typical case, we note that the fifty-two libraries of Buffalo and Erie Counties have been threatened, as well as the twenty-seven in Washington state and the sixty-five in Massachusetts; that all the library establishments in Spokane, Washington, have reduced their hours, as has also occurred in Denver; and that twenty of the forty-nine public libraries in Philadelphia must now operate *without librarians.* These librarians immediately wrote an open letter to their director, Elliot Shelkrot: "Your decision to select mediocrity over excellence . . ." prompting Shelkrot to respond to the press that this was the way it had to be, or else ten of the libraries would have to close their doors permanently.[2]

Today, the alert is everywhere: the citizens of Vanier rose up and sought to take over the management of their local library, which was part of three branches of the Ottawa Library threatened with extinction because of budget cuts; annexes from South Africa to Germany were given the stark choice between closing or reduced hours, whereas the freeze on future acquisitions was already confirmed. At the summit of the pyramid of the untouchables, the Lambeth Palace Library[3] in London was put in a shockingly hot seat: a commission of administrators and financiers of the Anglican Church broached the idea of moving it from its home on the banks of the Thames, where it had been host to the Archbishop of Canterbury since 1610—with its six hundred manuscripts, some dating back to the eleventh century, and twenty thousand printed works from before the year 1700, not to mention the rest of the treasures there—to a site that could not be any more ordinary and, worse yet, was located in the country. In the capital of liberalism, such a sublime location, costing £477,000 a year in operating costs, could not be reserved for an average of seven suffering scholars a day when it could be earning millions if it was turned into a tourist site.

*[The Salinas Public Libraries are still in operation as of the publication date (2009) of this book. —*Ed.*]

Furthermore, management did not take long to realize that the manuscripts intelligently displayed on the Lambeth website had received 4.5 million hits in 2004. Given the general outcry triggered in the tabloids by the first position they took, which was a bit too straightforward, the church commissioners then found it far preferable to knit together a voluminous report that demonstrated on its release in February 2005 that the only element missing, above the ancestral position of librarian and archivist, was director, about whom "it was expected that the salary required would be more than compensated by the gains in efficiency and improved services." In all the world's religions, this is known as attaining one's ends by moving crabwise.

The United Kingdom again trembled with indignation following the publication of "Who's in Charge?" the violent attack by Tim Coates against the deteriorating management of public reading, which in his opinion still provides indispensable support for the continued education of citizens. He was a former marketing director of WH Smith who subsequently became an activist London bookstore owner with the opening of his shop in 1992. This establishment was open twenty-four hours a day and included a café on its premises. Coates showed in his report, published in 2004, that lending libraries would be completely extinct by the year 2020 because the growth of expenses was inversely proportional to how frequently they were used. It so happens that the actual purchase of books represents only 9 percent of their budget against 55 percent for personnel, and for miserable spans of open hours: thirty-five establishments of Hampshire out of fifty-four were open forty hours a week and were poorly maintained, some even being in unsafe condition. Tim Coates therefore advocated a complete reorganization of the system, details of which can be found in his text available online on the Libri site,[4] a "charity" site for libraries in distressed circumstances. For example, he advocates simplifying the administrative procedures for book acquisition that effectively double their cost, and increasing the hours libraries are open by half again as much, or having these hours transferred in large part to evenings and Sundays instead of the current hours, which coincide with the times most people are unable to visit them. Consequentially, this man's nickname in the librarial milieu is "Ghengis Tim."

The budget cutters are not the only ones working against the library. They receive valuable assistance from the eulogists for efficiency and modernity, as we have recently observed symptomatically in Brussels.

In this latter affair, the Internet also gets the blame. The perspective offered by the digitization of books, although relatively vague and remote, has been invoked on several occasions by the technocrats, who are perhaps the same people information technology terrifies. Confronted by so much bad faith, one is led eventually to ask whether it would not be in the best interest of a good and beautiful library, in order to maintain its happy life, to live hidden from the computers.

In 2002, the Belgian Federal Administration acquired a "Copernicus Plan" for establishing efficiency in the ministries, which were transformed into SPF, Services publics fédéraux [Federal Public Departments], whose directors were no longer bureaucrats with lifetime employment but "top managers"—drawn from the private sphere to boot—and given six-year mandates. A "Kafka Plan" (yes) was launched at the same time for simplifying people's lives because "the administration must be transformed into a modern company." One of the first victims of this devastating creed was the Vesalius Library, where three thousand visitors annually from the Health Ministry and from the general public came to learn about the security of the food chain, tuberculosis, or the Marburg virus.[5] Slyly dismantled and transformed into a "public ticket booth" for the Eurostation, a station for the TGV,* its twenty-six thousand monographs remain homeless for the moment, whereas the considerable series of impeccable medical and scientific reviews dating from 1900 have already been clandestinely abstracted by other institutions, including universities, which may not be entirely legal. But while the collections of the library are noticeably absent from the future installation, where it

*[High-speed train. —*Trans.*]

will be necessary to order them in advance at best, one foresees a "closet museum" that will allow the most valuable books to reap "the attention they deserve." Deputies and senators invoked the traditional concerns, which drew a response from the minister that "it was not his job to keep the collections of books and magazines up to date. The library is a tool of interactive work with a living group,"[6] leaving the underlying impression that there is a strong possibility of its sudden miniaturization on the whim of the appropriate management and the idea those who govern have of the governed.

"A veritable gold mine of information for you," says a brand-new brochure about the libraries of the Federal Public Departments of Belgium, where, next to the Vesalius, stands in good stead the Quételet Collection, with its million volumes on economics, and the Foreign Affairs Library, wealthy with its considerable sources on Central Africa, as well as a good many other archives of the former ministries, all open to the public. As we have been seeing, this quality offers no guarantee at all that they will survive.

Is France protected from such concerns? Yes, of course, in the same way that such a spectacular concentration with dramatic effects in the press and in publishing was unthinkable only one year earlier. These are perhaps just the first drops of the coming torrent. All that can be seen for the moment are a few *small* closings; however, they are anything but small for their immediate victims.

And no fewer than three cases in three months.

The naive days of Carnegie are long gone, when the embryonic unions grumbled that the money spent on books would do much more good in the pockets of the workers, to which the boss responded that they would most likely just spend it on drink. During the time the BCE* of Renault Le Mans was disappearing, it was learned that the general management of the INSEE [Institut National de la Statistique et des Études Économiques/National Institute for Statistics and Economic

*[BCE stands for Bibliothèque de comité d'entreprise, library run by the works council, which is a shop-floor organization that represents the workers of a company. —*Trans.*]

Studies] intended to eliminate its library, keeping only the house publications. Not an echo of this appeared in the newspapers, in contrast to the affair of the library of the personnel of the City of Paris, which had also been condemned with its seventeen thousand volumes and forty positions, and where one bureaucrat of the city out of forty-three was inscribed. Its collections were "quite rich on the DOM-TOM.* It is necessary to make an appointment, by asking politely . . . that might work," testified the signatory Bidoc in June 2003 on the site volcreole.com after having read there *L'Enfant antillais en France* [The Antilles Child in France] by Jean Biarnes and Michèle Surhomme (Paris: Éditions L'Harmattan, 1983). In October 2002 the library had also organized "Pawòl an bouch, Kréyol-la ka vansé," a literary exhibition devoted to the writer and creolist Hector Poullet.

The article in *Le Monde*[7] on this programmed closing was relayed by the Turkish site Celiknet under the rubric Regions, which leaves the impression that the mayor of Paris will one day be regarded as the elected official of a commune far from Europe. Trimming down the budget cuts is not enough to hide the fact that the real problem, and the consequences that must then be suffered, is the spontaneous choice of public access to culture. In fact, unless the new device of the capital of France is not to become *cervesiam et circenses* [beer and circuses], wouldn't there be myriad other extravagant or ridiculous expenditures, such as the quays of the Seine playing *Les Sables d'Ozone* and the futile propaganda for the Olympics of small business in 2012, that deserve priority when it comes to what things should be reconsidered for a readjustment of choices?

The process that is dismantling or adversely affecting research libraries seems more hypocritical and larval: the document collections of the museums (the Museum of Man, of African Art, of Folk Traditions, and so forth), and the large public scientific establishments turned upside down by who knows what frenetic urgency are suffering enormously

*[DOM-TOM (Départments d'outre mer, territories d'outre mer), the overseas departments and territories of France. —*Trans.*]

because of it. Marie-Dominique Mouton: "Everything is happening as if these libraries . . . should abandon their own evolution, or accept the loss of their identity by melting down into structures organized in accordance with more generalist principles. In both cases, it is a part of the intellectual history of these past fifty years that is vanishing."[8] In all these cases, it is the denial of idealism that is being made the law.

In a parallel but more radical vein, if one might put it this way, another phenomenon is striking the small municipal libraries, beginning generally with their annexes, which are located, theoretically, in the neighborhoods where they are most needed. It could bring about the closing of all of them in quick succession.

The execration of the cultural site is at the point of becoming a new attitude in the hard sectors, where it can pass as an expression of the ruling authority. It has begun to show itself on the English side of the Channel, but for the moment there seems to be no watchman anywhere coming to any decision about this question of the future. The modest lending library of Shelfield, a branch of the Walsall Library, closed its doors permanently in March 2005 following repeated attacks from several yobs (a British slang word designating thugs who are on the young side), who seemed to have been excited rather than intimidated by the presence of watchmen, and who eventually set fire to the building. A bookmobile should soon be providing lending services, according to the town hall. In the meantime, readers are invited to visit the Municipal Libraries of Rushall or Pelsall. Like the rusted sign of the saloon in a ghost town, "Welcome to Shelfield Library" still greeted visitors to its homepage on the Web a month later, a page that led only to a broken link. There are (still) eighteen libraries in this county of the West Midlands, separated by several miles from Whitnash in Warwickshire.

On April 4, 2005, at 9:25 p.m., a fire was set in the library of Whitnash by "vandals" who introduced burning materials into the building through an air vent. Some young men wearing hooded sweatshirts are the most likely suspects, but although they were caught on film, their identity has not been discovered. The director of the regional libraries, John Crossling, confessed his disgust: "We are doing our best to provide a service to the local community and have got computers in there which are aimed at the younger generations. Yet it appears to be members of that generation who are behind this!" One brilliant adviser at Ten Downing Street suggested the government should prohibit wearing a baseball hat with visor and a hooded jacket at the same time. Beyond this, no one seemed willing to reflect on the true nature of this problem in Great Britain. So let's not think any more about it either; that way it can be a surprise.*

*And here it is: in only the time it took to send this manuscript to press, the library of Lewarde and the *mediathèque* of Auby went up in flames near Douai, just as they did in Boissy-Saint-Léger, Chalon-sur-Saône, Grenoble, and Saint-Étienne (November 6–10, 2005). The members of the government are completely absorbed now by reducing taxes on the wealthy as well as on the issue of secularism—this causes them to look at one another like so many herons and think in whispers: Should we be building suburbs in the country? Already some local elected officials are leaning toward the sole choice within reach of their imagination: remove free public reading from all towns that are poor and undereducated.

21

Paper Leaves by the Door and Comes Back through the Window

The apostles in favor of digitizing everything can rightfully rejoice at the announcement of all these appalling events, so long as they demonstrate that the placement of books online for personal monitor screens in a snug apartment represents the sole, rational, imaginable, and safe solution. We can support this idea, on the condition, obviously, that enjoying works that are in the public domain remains free everywhere and for all time.

Considering what Brewster Kahle has already accomplished—for example, he donated one petabyte of the old sites archived on his Internet Archive to the Bibliotheca Alexandrina, and contributed to the Million Book Project as well as to Open Content Alliance—he is most certainly a man on whom the future can count. We now owe him for the bookmobile (which he did not trademark: we are entering the sphere of free access).

The bookmobile is a van, a bus, a delivery tricycle if need be, containing a laptop with a satellite connection to digitized libraries, one or more connected printers, and a binder. The total cost of the equipment is less than three thousand dollars. In ten to twenty minutes, and for the price of one or two dollars, basically the cost of the paper, you can continue on your way with a public-domain book, most often one long out of print or one that cannot be found by any means where you live. *The Adventures of Alice in Wonderland* is apparently the most requested title. The idea was first triggered by the fact that at Harvard Library, the cost of extracting a book from the shelves, entrusting it to the reader, then placing it back on the shelves after it was returned amounted to more than two dollars. Therefore, giving away the book is more entertaining and less expensive. Hunting down the book online and printing it is child's play, so the bookmobile has enjoyed wild success in India, in Sabah (Malaysia), and in other spots around the world such as Kenya, where, it is said, it travels by camelback.

QUOTE OF THE DAY

By 2047, libraries offering free public access to scholarly literature were a dim memory.

RICHARD STALLMAN, *THE RIGHT TO READ**

*A short and most remarkable text published in 1997 and available to read online with a note from the author, updated in 2007, at www.gnu.org/philosophy.right-to-read.html.

22

Library, Arise!

For Benjamin Franklin, there was no doubt whatsoever that a library should share and lend books to its subscribers. From 1731, he did all in his power to make this a reality. He would state this wish in a very charming way in his autobiography:

> These libraries have improved the general conversation of the Americans, made the common tradesmen and farmers as intelligent as most gentlemen from other countries, and perhaps have contributed in some degree to the stand so generally made throughout the colonies in defense of their privileges.[1]

For Victor Hugo, "The book, as a book, belongs to the author, but as thought it belongs—this word is not too overreaching—to the human race. Every mind has a right to it. If one of the two rights, the right of the author and the right of the human mind, need be sacrificed, it would certainly be the right of the writer, because the public interest is our sole concern, and everyone, I have no hesitation declaring, should pass before us."[2]

But for the Universal Declaration of Human Rights in 1948, the matter became a bit more complicated. Article 27 states:

1. Everyone has the right freely to participate in the cultural life of the community, to enjoy the arts and to share in scientific advancement and its benefits.
2. Everyone has the right to the protection of the moral and material interest resulting from any scientific, literary, or artistic production of which he is the author.

Out of this contradiction in Article 27 would suddenly emerge, a half-century after its hasty formulation, a fissure between two worlds that would only grow larger with time: the ideal and the venal.

This has triggered a world crisis of "the government of learning," whose efforts aim for an ever-greater enrichment of private interests to the detriment of the free and open propagation of knowledge as it has always existed. The prevailing trend is a "lockdown of the public domain"* that appears all the more bloodthirsty inasmuch as the fields concerned are the fundamental ones: from health (I am thinking primarily of the patents on AIDS medications, which prohibit the poor countries, generally those most heavily affected, from having access to treatment) to the simplest software applications—which are also the most commonly used by everyone, including by the administration—that a skillfully manipulated Europe tried in 2005 to make patentable.† In the libertarian front of "Scientopia,"‡ for those who share from the onset the belief that scientific innovation is the sole guarantee of absence of pressure from the private sector, let's note in passing a titillating generality by Jean-Claude Guédon, professor of comparative literature at

*Florent Latrive, *Du bon usage de la piraterie* [On the Good Use of Piracy], preface by Lawrence Lessig (Paris: Exils, 2004). One finds in this precursory work, sporting a somewhat inapt title ("On the Poor Use of Copyright" would have been closer to the actual contents), in addition to the terrifying information and reassuring exits offered, that public domain with a price tag has existed on several occasions in France.

†For the moment these efforts have been unsuccessful: see the positions of the associations advocating free software (free for everyone once its development costs have been paid for, in contrast to the patented products that will require eternal remuneration): ADULLACT, AFUL, APRIL, and FSF France, at http://these.univ-lyon2.fr/.

‡[An Internet term that pushes scientific solutions for the world's ills. —*Trans.*]

the University of Montreal: only 20 percent of the planet's inhabitants profit from good distribution of the world's available knowledge. Humanity is therefore deprived of the 80 percent of potential contributions represented by the nonirrigated brains.

Do you hear some awful racket? It is the world tossed to the dogs: hordes of quibblers and disputers hunt down every pocketing of legal profit that the social body might reveal in its various palpitations. We seem to be entering an era in which everything can be combed through for "pickings." When a plane crashes, there are suddenly more lawyers than nurses. A new corporation is born, whose device could be, quoting a journalist: "Everything has its price, everything is litigated."[3]

Hovering over our little lives is the ever-growing shadow of WIPO, which serves as a rake for the WTO [World Trade Organization]. This World Intellectual Property Organization is "dedicated to promoting the use and protection of works of the human spirit. These works—intellectual property—are expanding the bounds of science and technology and enriching the world of the arts. Through its work, WIPO plays an important role in enhancing the quality and enjoyment of life as well as creating real wealth for nations." Using the same tone used by papermakers when proclaiming themselves the saviors of the forest, this is the chant found on the site for the U.N. authority WIPO (whose English acronym sounds like the brand name for a dusting product).[4]

Warning to adolescents: "Intellectual property and you. Did you know that intellectual property is never far from you? The things you use in your daily student life—the clothes you wear, the books that fill your backpack, or the music you listen to—are all concentrated forms of intellectual property. You may not be aware of it yet, but intellectual property is everywhere in your world." With these words, the terrified kid feels his world cracking apart in front of him; unless he opts for flight, he will quickly grasp that confronted by the environment he's been promised that is so heavily padlocked and heavy with threats, one sole means of reaction is left to him: to turn into a hooligan.

Young or not so young, the physical individual is not the only one suspected of naturally having bad intentions. We now have the befuddling case of the deserving library of Lisieux,[5] now held hostage by its success as the digital pioneer in France.

"Subsequent to a demand by a company overseeing author copyrights and for the purpose of not infringing upon the patrimonial rights of authors' potential heirs, in application of the articles L. 123–1 new, 123–8, 123–9, and 123–10 of the intellectual property code, a certain number of authors' texts, wrongly attributed to the public domain, have been pulled from the catalog of the electronic library of Lisieux. We will henceforth apply a principle of systematic [and somewhat stupid] precaution in the selection of the authors and texts we wish to present on the site. Only those authors will be selected who are believed to have been dead for 86 years [(70 years + 1 year = art. L. 123–1) + (6 years + XXX days = art. L. 123–8) + (8 years + XXX days = art. L. 123–9)]." Following this message of contrition is a demoralizing list of dusty but amiable texts, likely excluded forever from the traditional circuits, signed Bloy, Beauclair, Cim, and so on.

23

Striped Uniforms

I find it scandalous that the SNE may not be the administrator of an establishment [the BnF] that could not function without publishers.* . . . Thirty years since we let ourselves be duped with the photocopy . . . it is in our best interest to make the first move, to put locks and counters on the right places." This barking of the jailhouse turnkey betrays the ambience of the discussions that the National Library, a good daughter, soon engaged in with the publishers when it began revolving around the PLAO and wished to digitize the texts still under copyright. These discussions lasted for more than ten years, mountains of registered letters were exchanged with each of the companies concerned, internal memos were accumulated, thousands of hours of meetings were held, all for a half-starved outcome: in exchange for an arbitrarily calculated licensing fee paid by the institution each year, 3,558 titles can today and until December 31, 2008, be read on-site on the screen by more than one reader at a time.

*An obviously thoughtless assertion, to the extent that not only does this archive pay real money for the books it acquires in addition to serving the publishing industry as legal deposit, but also, in extremis, because the library would also fulfill almost the same role with just those books that have long been out of print—that fell from the mouth of the president of the Syndicat national de l'édition during an interview with *Le Figaro* in 1994.

One of these publishing houses, long thought to be a paragon of textual discovery, recently experienced a retreat toward guaranteed value as well as added value: a balm for the banker, but an action accompanied by a posture that looks quite cramped when facing the future, with a regrettable effect on its image. It nonetheless remains a model in this regard for the majority of its colleagues. Therefore, when the book trade sought to smoke the electronic peace pipe with the National Library in 1997 and "to mend a painful past,"* although the latter had admitted from the outset that it had emerged from this quarrel with its fingers burned, the publisher in question and one or two others spent years dragging their feet and finally *never* signed the protocol. In a similar vein, questions have also been raised about the severity of the punishment it imposed on a website guilty of having reviewed too heatedly, and citing entire paragraphs—and likely selling more copies, but the publisher claims to be convinced that the opposite is true—one of the cash cows in his portfolio.† It is likely that if you walk down certain tiny streets of the Saint-Germain-des-Près quarter and shout "Library" or "Google Print," buckets of dirty water will be poured down upon you from the machicolations.

While still imagining itself to be one of the wonders of the world, a large portion of French publishing is no longer anything but a provincial lighthouse. Like its famous counterpart at ancient Alexandria, it should take into account the fact that it is inflexibility during earthquakes that causes the greatest damage.

---

*This note, in the resolutely optimistic hand of a woman director of the BnF, appears on the margin of an internal memo reporting on the bargaining. BnF archives have now been made fairly easy for researchers to consult.

†The cream of the legal profession having been noisily called to the rescue, the story of these mediocre vicissitudes is fairly easy to find on the Web. Guy Laflèche, professor of French studies at the University of Montreal, has put together a panorama of literary sites dating from before 1996. The result smacks of both the cemetery and a minefield: www.mypageweb.umontreal.ca/lafleche/co/pr4.html.

24

Shrouds

Blood rushes to the forehead when one is confronted by obtuse stupidity: clutched around the throat by legalism, one sees the avant-garde of the virtual libraries laboriously competing with each other, offering the same minor masters from the end of the nineteenth century who died on the right side of the frontier of freedom. By this token, the fairly wretched Guy de Maupassant has become the author with the greatest online presence, with a narrow selection of novels that many readers know almost by heart and who often own, as their price is not too high, editions from his era, although—and this is the case with many young amateurs with a hearty appetite—they have trouble appreciating them, given the uncustomary, if not to say downright obscure, literary style and vocabulary. The defenders of the Great Digitization find their ultimate argument of a somewhat ontological quality here: what rational mind from the future will consider without laughing that through the concomitant emergence of a copyright fundamentalism, the most formidable technological discovery of the time could be used only during the first decade of its existence as distributor of texts written in a language that three-fourths of the population no longer comprehended? Rarely in history has a discovery been seen to develop its applications inside such a gummy ineptitude.

Here is some glue for more problems in the meantime. Every book tips—some even say "falls," like a ripe fruit, although nothing prevents one from claiming like Hugo that it rises—into the public domain at varying times depending on the country. In France, it is sixty-nine years after the death of the author, plus the years of the war, which count double, and an additional thirty if the author "died for France." A book can have been published in 1900 by a precocious genius (let's say he was then twenty years old; once upon a time things like this were quite common); if he died a centenarian, the title would not be free of copyright until 2066. However, the generally accepted time span is fifty years, according to the old Bern Convention, but contradictory tendencies are in the midst of pulling legislators this way and that, for example, in the United States. Some, like the nonnegligible law professor—we cannot get away from it—from Stanford, Lawrence Lessig, would like Congress to impose a copyright that would be renewable every five years so living proof could be provided that a copyright holder exists; whereas others, like the shareholders of Walt Disney, are apparently applying pressure to have the term of copyright extended twenty years every time the former term is set to expire.

Eighty-six years! One hundred twenty years! The current profusion of legalistic entities would have made Victor Hugo giddy, but one can imagine the howl that would have been elicited from him by this extension grafted onto author copyright that now only anonymous management companies benefit from. The actual heirs and even the publishers crumbled into dust ages ago, and companies do not even pretend to hunt down the rights holders, as this activity would entail a disproportionate expense. It is a universal pimping system erecting its big top before our very eyes.

Pushing the length of time commissions are paid out on these titles back down until twenty-five years after the death of an author would be much more reasonable and make it so that those who remembered him or her, such as grandchildren or a great-nephew, would still get a few Christmas presents out of it. Next it would be necessary to return to the system of "the public domain for which you have to pay" that Hugo extolled, rightfully, in these terms to his stupefied colleagues: "Do you know anything prettier than this: all books that have no direct heir fall into the public domain for which one has to pay, and the product serves to encourage, invigorate, and fertilize young minds!" How he would rail today against the cultural administrator who, training himself to speak like Louis XIV, especially distinguishes himself at the sides of the multinationals championing the works of hip-hop spirit?

In France, the miserable protest libraries' right to lend books instilled by Brussels contributes to their impoverishment and will perhaps one day be seen as the precursory sign of their rarefaction, if not their disappearance. The publisher Jérôme Lindon will go down in history for having noisily associated his name with this sleazy maneuver. An unrelenting enemy of libraries whose contempt "leaked through the telephone"[1] produced, among other things, a famous and premonitory article[2] whose argument, without changing a line (except for the untruth about the mediocrity of the public reading offered over the last century), could be reused to require either "Hachette and Vivendi" or the television channels to transfer 10 percent of its profits to the management of the book, in order to finance the right to loan books! Conversely, the French government defused the problem as in a hostage-taking through a financial pirouette to the detriment of the libraries and in a way that the public, therefore the media, did not really pay any attention.[3] Old writers cried victory, but they could not have been any more mistaken. This measure will fatten publishers' coffers and be literature's comeuppance: the Sofia—Société française des intérêts des auteurs de l'écrit—is a creature of the Literary Guild associated with those publishers who are most aggressive at perceiving the funds disbursed by the state in proportion to library members (twelve million euros in 2005) plus a 6 percent tax on the purchase of books for public reading, which has a direct and detrimental effect upon the revenues available for that public reading. Still too young to answer the questions of the old gentlemen (its approval was due March 14, 2005), Sofia cannot say how much money it has in its account, nor what it can plan on transferring to publishers, and, as a secondary consideration, to the complementary retirement accounts for writers. The latter do not recall having been consulted to any great extent during this operation and, by a large majority, do not expect to realize any great profits—unless, apparently, they are already authors of bestsellers.

We should not think, though, that everyone would be in favor of this. In addition to the 327 of their number who signed the petition opposing this, Baptiste-Marrey, François Bon, Daniel Pennac, Jean-Marie Laclavetine, and Michel Onfray signed a text entitled "Prêter (un livre) n'est pas voler (son auteur)" [Lending a book is not stealing from its author],[4] a gravely offended stance that refutes the twisted choice of the supervisory authority and which deserves an updated reprinting, the file remaining a litigious one. It is regrettable that the *Bulletin des bibliothèques de France* did not pursue its 2000 investigation on the state of the right to lend in other countries besides France. This periodical does inform us in its issue no. 45/2 that Spain has completely exempted all public libraries from this tax: The land of Quixote, champion of the *vergüenza.**

The library is a place where the best customers for the bookstore are formed. A strategic error is therefore added to what is a major scandal, one that disputes two millennia of freedom for people to visit premises desired, financed, and maintained by the will of the prince or the community, for the purpose of reading, without paying, the books that they would obviously not have purchased. What devil has gotten into us, then, to attack such an ancient and excellent principle, one that is indisputably beneficial for the individual as well as for the society?

But the most appalling thing was yet to come.

*[*Vergüenza* means "the sense of honor and dignity." —*Trans.*]

25

Purse Strings and Police Cordons

While knitting its collective legal brow, the European Parliament also enjoined its member states to apply its coordination directive 2001/29/CE* with greater ardor. For the archetypal commissioner Fritz Bolkenstein, this would be a "decisive step forward," one that would first and foremost impose "an elevated level of protection" on author copyright, "as these rights are essential for intellectual creation."

Now, given the tiny island that literature represents in the world market, these terms are quite obviously unacceptable, because what is being prepared for us is clearly the opposite of what is being claimed.

I will abstain from pronouncing on whether the majority of audiovisual productions exist only by virtue of money and for the purpose

*Or DADVSI, short for the project of the law relating to author copyright and related rights in the information society (and not "on the Internet," as can be read on more than one blog). All the current news will be found updated on a site fomented by associations of archivists and librarians who have tried in vain, despite the 2,800 signatures on their petition, to have this renegotiated. See: http://droitauteur.levillage.org. The action was vigorously relayed to the public (100,000 signatures in several days) by FSF, the Free Software Foundation, www.eucd.info. Finally, *Le Monde* devoted a lucid but tardy extended report on this subject, now ten or eleven years old, in its November 22, 2005, supplement "Economie." We know what happened next.

of making more of it, but I have no hesitation maintaining that putting them under the heading of "intellectual creation" is a snake in the grass that even those most ready to accept all kinds of cuisine could not swallow. It is glaringly apparent that it is high time to organize a summit on a neutral terrain with the petty speculators and crooks of all kinds in order to examine the matter from every angle and to share together in an irrevocable manner the vocabulary currently used by everyone, even cabinet ministers, as is the routine in a friendly divorce settlement.*

Furthermore, bringing the literary invention into line with such misuse of language is a sign of crass or biased ignorance of the history of writing, the majority of whose books have hatched in spite of the most elementary economic good sense. Can you imagine Lautréamont doing the math to see if he could buy a villa on the Basque coast? Don't the European Community leaders in Strasbourg know that Julien Gracq had to turn his savings over to José Corti in order to get the latter to give his work even a glance? A hundred recent examples of this kind could be listed, a thousand if you count China. This is because writing is always for free. If marking up a piece of paper every day sometimes generates what's needed to pay the exorbitant rents of Paris or New York, it results, from the viewpoint of any writer worthy of the name, either from his personal pugnacity or from a miraculous coincidence but never, ever, even for those who make a handsome living from their writing, from those funds that are deemed "essential for creation." Few beneficiaries of a fat literary prize have it in them to write a good book *afterward,*

*The brand-new label "intellectual creation" knows no limits when it comes to excessive interpretation and misuse. Following the classification of the broadcast "Pop Stars" as an *artwork* so it could benefit from a grant from the National Cinematography Center, one or two elder film directors, looking out of place on the small screen, protested (but once you have sold your soul to the devil, you should not be surprised when he next demands the shirt off your back): "A television channel is not an author, no more than a production company is. What kinds of crimes will profit from this confusion of genres?" Jean-Louis Comolli, "La creation audiovisuelle tuée par les siens?" [Audiovisual creation killed by its own?], *Le Monde,* June 1, 2005.

as the most invincible motivation for writing is the eagerness to be, if not known, at least recognized. This is, moreover, why well-established novelists gather once a year to decide which of the newcomers shows the most promise of becoming a great writer and overshadowing their own work. They then single him or her out as the laureate, certain that this will transform the budding prodigy into a mere hack.

The authentic writer thinks only of his next book; he leaves to others the concern of looking out for his interests, with the risk that they sometimes swindle him: his publisher or publishers, his agent or agents, all of whom are shielded by the offices of powerful attorneys. Those who brandish "author's copyright" at every turn of phrase are therefore displaying blatant hypocrisy. It would be more correct for them to say: "the right of the copyright holders of which the author is one among several." Proudhon's definition of property* is again shown to be most accurate. Gathered into professional groups, bodies, and unions, the *nomenklatura* that lives off the intellectual work of the authors has exclusive representation of literature when a parliamentary commission holds hearings. For example, among the seven hundred personalities questioned by the spokesman for the committee on the law on the "author and associated copyrights" while it was being drafted in June 2005, not a single author or book historian was asked his opinion, and of course not a single library user.

Both going out of one's mind and improving it by virtue of books being made rapidly available to one stem from a choice that forms an ancestral ornament of democracy, and the relentless defenders of copyright still display some hesitation when it comes to making a frontal assault. This explains the embarrassed and procrastinating attitude displayed by governments. But the dematerialization of the text indiscriminately kicks open all kinds of doors both to the imagination and to the bailiffs. Digitization, after all, is digitization! The tragedy can, therefore, be summarized as the pure "contamination of the concept of

*["Property is theft." —*Trans.*]

author copyright by the concept of patent." Buying is no longer enough; one must pay to read. But, "starting from the moment when reading can no longer be freely shared, it is the fundamental situation and primary role of culture that is being challenged. Accepting the placement of reading under the clutches of the mercantile economy fatally accelerates the reduction of cultural productions in favor of entertainment and mass production, which alone would be made available to all thanks to being sponsored by its advertisers."[1] This no-nonsense article by Christian Vandendorpe has lost nothing of its acuteness except for the fact that what was conditional in 2001 should now be changed to future tense, for it is now here in the present. With respect to libraries, in any case, Richard Stallman's prediction has come true forty years in advance.

A book, or the majority of books known by literary history, as an object to be kept and revisited to discover new meanings does not have much in common with these productions for common consumption that have become increasingly quick to appear and have now reached the point where they are immediately forgotten once they have been seen or heard. Nothing is more comical, moreover, than the indignation of the rights holders who claim to have been deceived by the very individuals they have targeted as victims of a shameless mugging. If downloading is a crime, then profiting from disarmed minds is an even more serious crime. Who, on the other hand, can be compelled to believe that the literature junkie will go to the trouble of pirating the latest biography of Alexander Dumas?[2] And what, by all great gods, is he going to do with a pile of 710 A4-size pages—use them to raise his bedside lamp higher? It is significant that the sole publisher in the world who openly, immediately, and violently rose up against the Google Print proposal was the publisher of the Harry Potter books, Bloomsbury, whereas the first to hurl themselves into the cyber melee are located on the other end of the intellectual landscape: Oxford University Press, Cambridge University Press, De Boeck, and Éditions de l'éclat in Paris. Some publishers have truly gotten it: the Internet offers arms as well as armor;

a moral renewal may well be opening there, whereas the profiteers are hunting down the friends of the book while brandishing their horrible straitjacket.

With a lot of fuss, they are suddenly raising police cordons around the BD,* hip-hop, and the scantily clad cinema—which implies tight control over the electronic behavior of every individual: FSF† describes this as a "technico-totalitarian shift" of the right. The legist swipes everything, library and literature included. Only the fear of being fooled seems to guide him.

*[*Bande dessiné,* or comic book. —*Trans.*]

†[Free Software Foundation. —*Trans.*]

26

Smocks

Confronted by so many hopeless mix-ups and the indigestible nature of the European Union coordination projects, the authorities during this time frame (April 2005) ordered Tolbiac survivor and state counselor François Stasse to turn in a report "on the access to the digitized books held by the public libraries" for the purpose of the semblance of a discussion by the deputies and the adoption of said law that was scheduled to come up for a vote on June 7, but which an electoral hiccup caused to be adjourned—without lessening any of the danger it represented. What did this Mr. Stasse have to say? Specifically, that it was a question "of enacting a principle of forbidding access to digitized books under copyright, whereas the possibility of gaining access would be given the status of the exception. This relationship between the principle and the exception is the reverse of the one that has prevailed up to the present in the Gutenberg paradigm, whose principle was freedom of access to the printed work and in which interdiction was only the exception." Having elegantly stated that henceforth everything would be "less good than before and that it was our duty to make it so," out of the blue, the commander's finger pointed to the countless number of poor books hidden away in the shadow of the catalogs, the books spat back out by the sales circuits—the purgatory of authors purged by their own publishers. These are the books that the law governing author copyright would prevent, nonetheless, from being reproduced either for the purposes of conservation or for being transformed into a form that can be read online—books that any of us can open every day inside the library. The spokesman promptly baptizes this shelf as *illico* the "gray zone," gray like shrinking light, gray like the smock of the school prefect or grocer of yesteryear, unaware that this pejorative color is already being used to designate an entire body,* but a little additional ambiguity is no cause for fright in an outrageously confused case file.

A motherless posterity, these objects forgotten by commerce form the gigantic and covetable capital of the large libraries, where they are caught in the trap of both juridism and the hazards of commercialization. In the United States, Brewster Kahle (the pioneer of Internet Archives) was the first to reveal this significant treasure, which he estimated at thirty million titles. He also stated that researching the most recent right-holders would already be too expensive for the copyright collection societies. Google-or-another-Incorporated grasped this easily and hopes to get its hands on the texts that fall under the heading of gray literature as soon as possible for nothing. The French language has a pretty way of describing this in one word, something that requires a long phrase in English: *carambouille,* "the quick turnover of goods that one has acquired unlawfully."

Inasmuch as they have not been asked, the experts have refrained from mentioning how this dematerialization of millions of printed books and texts would be financed (although if we consider the most recent past, publishers are holding what in any case would be the beginning of an answer: they are in possession of the electronic files for every book). On the other hand, they are venturing to assert, "The question

*"'Gray literature' designates all the documentary material produced for a limited audience outside of commercial circuits of publication and dissemination and on the fringes of standard bibliographical indexing arrangements" (AFNOR) [Association française de normalisation]. Examples: working or research papers, conference proceedings, theses, and so forth. It will never be too soon for the word *literature* to be placed outside this entire *zone* . . .

is knowing if this double usage can concern the same books at the same time. The response provided by the law to this question is currently no. One must in fact await the author's rights expiration; in other words, as a general rule, wait seventy years so that use of the digitized form can succeed the use of the book printed on the support of paper," a purely gratuitous, personal, and barely consistent interpretation—the "in fact" attempting to pass itself off as unique and sufficient proof, with "succeed" then implying that it was a question of the two supports being used simultaneously in the preceding phrase—of a law that is vague on this point. As a consequence of the apparently exaggerated emphasis, the text aligns itself with the religion of the jailers: a reader would henceforth have to pay for reading, which could be carried out only at a fixed and secured post, installed at maximum in ten establishments of this county—in other words, the very opposite of the service digitization would naturally tend to fulfill.

No one knows if this idea will catch hold of anyone's attention,[1] if only because of its unusual nature: Can you imagine the member of a library benefiting as ever from a paper copy of the book a mere yard away from someone who has paid to decipher it on a screen surrounded by virtual barbed wire? Or will all the books be withdrawn from lending as they are digitized, as is often the case in a sneaky and often unjustified manner with microforms? The next perspective that inevitably looms would then be to purely and simply shut down the premises.

But wouldn't this be precisely the great secret objective? The desire in high places, or in certain low recesses, to eliminate the guardrail represented by the Higher Council of Libraries (CSB)—to which no president has been named since 1999—a desire half-avowed in Grenoble in 2005 before a conference of librarians flabbergasted by the affront—could foreshadow the scathing admission.

The presidential whim for a new kind of library in 1988 has, as we have seen, degenerated beneath the mob of social climbers into a convulsive removal of all the books of the BN [Bibliothèque Nationale] and the surprise parturition of the BnF. This was already a political abduction perpetrated upon a large and admirable institution whose weakness and glory are never to raise a fuss. The free-for-all has now been forgotten, but it did leave scars. Two years before his death, Jean Gattégno knew that history would be remorseless, and he tried to arrange matters so that his name would be remembered more for his expertise on Oscar Wilde and Lewis Carroll than for his role as an accomplice to a national exaction. Would this old-fashioned dignitary today see confirmation of his reservations in the shameless plundering that was triggered?

Let's be serious: these millions* of the clearly living dead have been handed over to whomever politely asks for them from the date of their publication. And they should continue to be handed over unless the decision is made that public reading is henceforth unconstitutional. What we are waiting for at this juncture is for the reading public and the taxpayers to bestir themselves and make their resolve heard to preserve this fundamental right. Next, imagining that a magic wand had digitized their collections for them, the libraries should declare that it is their understanding that they should also offer these works for consultation from a distance, as their mission enjoins them to. This reflects their thinking, if not what they stated outright, in 1994: "Digitization and the right of consultation, whatever their modalities, simple or endowed with other functional features, should be free in conformance with the customs of consulting other printed material."† As a nonprofit entity that exists only to serve the citizens, the library should not be taxed for something it gives away as a gift.

The matter that should now be brought up for consideration at the highest level is: Do we accept the historic responsibility for having ceased to supply assistance to our people trying to educate themselves? The response to this simple question is equally simple: As there is no

*The archive was unable to answer a request for precise figures. François Strasse, who was director from 1998 to 2001, reckons 2.5 million titles. It could quite easily be twice that number.

†The consensual opinion reached by a December 16, 1994, meeting in preparation for upcoming publisher negotiations (BnF Archives).

lack of dirty money (the lottery, armaments, television . . .), the state can launder some of its wheelbarrows-full of billions to pay in a transparent fashion the additional and completely laughable sum asked by the publishers pulling their authors' strings, and, if it still proves necessary to reduce the expenditure, without the first cuts being directed at the books they no longer sell, some 80 percent of the total, so that the public can read both on and off the premises everything contained in the archived writings.

But it so happens that in its craving for less complex questions, or with an eye to getting the bitter pill swallowed by including some chocolate Christmas truffles—depending on the points of view—the Bourbon Palace [home of the National Assembly] was ordered, after the debate had been delayed for months and without waiting for all the dubious evaluation of the counselor to be assimilated, to declare a priori irrevocable twenty out of the twenty-one exceptions proposed by the European Parliament. Under the guise of helping them out of a tight spot by adding some flexibility to the situation, Parliament was doing so only in order to get a better shot at destroying the balance established in 1948.

It was a dual performance of dupes that served to its impassioned audience the strange troupe in which no one, except the male committee spokesperson assisted by the minister, played the roles he or she had rehearsed. The defenders of freedom crowed at the sight of order's customary friends voting with them for the establishment of an "optional legal license" for downloading, a timeless invention that is as unrealistic as it is unrealizable. The details were spelled out in the newspapers of the time, and we will not revisit them, but the most scandalous moment went unperceived. This took place while the session of Tuesday evening, December 21, 2005, was getting off to a lifeless start:

> Jean-Luc Warsmann—My amendment 111 is intended to add three additional paragraphs to Article L. 122–5 of the intellectual property code, to target respectively the acts of specific reproduction performed by libraries open to the public, by teaching establishments or

by museums, or by archives, which seek no commercial or economic advantage, either directly or indirectly.

The Committee Spokesman—I am against this. We have decided to retain only one sole exception: that which concerns the handicapped.

The Minister—I am of the same opinion.

Patrick Bloche—Like he does on all essential questions, the minister sends us back to the contractual context. It so happens that we are legislating! Our duty is to respond through the law to the concerns of librarians, teachers, and researchers.

Martine Billard—I too support this amendment, and the minister's response is unsatisfying. It is libraries of all kinds—from municipal libraries to company libraries—that are forced to negotiate, inch by inch, to obtain the right to use digitized books. . . . It is high time technological use is taken into account, and to authorize this use in libraries without waiting for long and difficult negations between quite different actors to take place. It would therefore be sensible to include all public and incorporated libraries in the law. We would thereby encourage frequenting of these establishments and access to knowledge!

Christian Paul—I do not know if we will be making historic decisions on all subjects, but in this vein, it is necessary, with the amendment of Mr. Warsmann, to create a cultural exception to answer the concerns of hundreds of thousands of professionals.

By a majority vote of forty against twenty, amendment 111 was rejected in less than a minute. The next morning, librarians scratched their chins and discovered with stupefaction how the future of their profession had been mowed down and trampled underfoot by legislators mobilized to defend the revenues of Johnny Hallyday* and his parasites and shareholders.

At least there was the semblance of a debate—which constituted

*[Johnny Hallyday is a French superstar, known primarily for his music but also as an actor. —*Trans.*]

for the majority a sufficient victory—one that opposed the party of the "riposte" to that of the "license," retrograde positions clashing equally with the evolution under way. This is why the real readers of real books are at risk of joining the Internauts in their revolt against the false problem of downloading. Here was additional proof that neoliberalism's belly is crawling with the larvae of a new-look obscurantism, in which laws and trends compete in mercantile narrow-mindedness, in defiance of the other and with the gesticulation of cudgels, as if the "information society" regards open-mindedness as public enemy number one.

The French—and only the French—therefore zealously opted for total prohibition, the new expression of a subtlety reasonably well handicapped for one or two generations.

Will free and unobstructed public reading end up being pulped because of a fantasy about piracy that is no threat to the book? Will the digitization boom that should have been entirely at its service instead serve to bump off the very definition of the library, its reason for existence, and everything on which it is founded? The arrival of digitized books on the scene is thus obviously no minor phenomenon of the intellectual, philosophical, and moral plane, but one that concerns everyone; its very nature deserves that everyone stop and take a close look at it. But we retain the impression that legislators,* caught short by a feverish future and quelled by their favorite lobbyists, have no fear whatsoever of being mistaken for gravediggers under the full moon.

---

*Immunity does not equal anonymity. The UMP Deputy of the Nord Department, born in 1947 in Tourcoing, Christian Vanneste supports the restoration of the death penalty and is also the author of the famous paragraph on the "positive aspect" of colonialism that was voted on in 2005. His interventions and political commitments are amply detailed on the captivating site of the National Assembly, from which nothing or no one escapes, but which hardly explains by virtue of what vice the individual chosen as the spokesman for a law as fundamental to the mind as that on the DADVSI is an elected official whose notoriety seems primarily based on the monotony of his homophobic statements. www.assembleenationale,fr/12/tribun/ficheid/2875.asp [The UMP (Union pour un mouvement populaire) is a center-right party founded in 2002 and currently the majority party in France. —*Trans.*]

Let's dream for a moment: with Europe imposing a detestable law upon us, shouldn't we eliminate our National Assembly if all it can do is aggravate matters (and while we are at it, why not do away with ministries and other redundancies, as they did in Belgium)? This might be easier than getting rid of the European Parliament . . .

But inasmuch as the elected officials want repression, they may perhaps end up with a real brawl on their hands. "Tomorrow, will we need to place a lawyer next to the lending desk or the consultation terminals?" the circumspect Association of French Librarians began to wonder, as it scrutinized these dark clouds at its Grenoble conference in 2005. In a June 20 motion concerning the European directive, the librarians hugged the walls a little less and "again condemn with the greatest resolve the disastrous imbalance that this plan risks establishing between producers and users in the nation. It is the desire of the ABF [Association des bibliothécaires français] that the principles that governed library operations on the world of paper yesterday, principles guaranteeing citizens free access to information, will be preserved in the digitized world." Now that is what was said. But even if it were written, wouldn't it be necessary, given the wretched esteem with which the legislative worthies regard this profession, to go much, much further?

In June 2006, for its one hundredth anniversary, the ABF gathers at the Porte de Versailles.* Suddenly shouting erupts: "Enough motions, we want action!" The conventioneers spread out into the streets and march upon the rue de Valois† brandishing pitchforks . . . A charming vision . . . And what if this large body that the caption describes as sensible as an *image d'Epinal* [stereotypical image], but made up of people rather hostile to authority when it is in the hands of profiteers and practicing Catholics, was to one day shake off its reservations to wave its guardianship in the eyes of all?

Against directives turned into diktats and other assaults, on the

*[The area where one of the main convention centers of Paris is located. —*Trans.*]

†[The site of the Palais Royal, home to the French Ministry of Culture and Communication. —*Trans.*]

right to learn who is using a formidable technological advance as their pretext, against an embryonic ban on the nonprofitable, we see here the ideal librarian, die-hard defender of author's copyright and everything else that one would like to commercialize—provided it in no way blocks, like the well-known definition of liberty, the intrinsic and untouchable mission of the library.* It will choose this motto for itself: *I provide food for thought.* This will be short enough for the banderoles.

Drum roll, please: The seizure of the written word constitutes the most formidable burglary of all time. That some Californian descendant of Fantomas† made the attempt was only to be expected. That the French Republic is surreptitiously donning its own burglar's mask should be vigorously denounced and attacked.

The public has yet to say anything. Yet it is the public that has been the most greatly wronged.

Librarians and readers, it is the same combat.‡

*This is what Yves Alix (Bibliothèques de la ville de Paris) has basically suggested. And the professional code of ethics published on the ABF site on March 23, 2003, does not say the opposite: "The librarian applies the policy of his tutelage when it does not contravene the general laws, the eternal and specific missions of the establishment, as well as the values defined in this code."

†[Fantomas is the name of the master criminal who was the star of French pulp thrillers published in the early decades of the twentieth century, a Gallic counterpart of Sherlock Holmes's archnemesis, Dr. Moriarty. —*Trans.*]

‡The appeal doesn't seem to have been heard by all: from March 2008 the BnF has agreed to launch a program of advertising for commercial aggregators of copyright books. As said on almost every page of this book, free public reading is at its agony.

27

Against the Grain

One library closes, one library opens; it is a growing imbalance. The one that closed was the last chance for our suburbs tumbling back into the Pleistocene Era to read a book; the other is an architectural show-off whose magazines have sparkling pages, a symbol, a repeated political gesture.

Let's look, for example, at the Seattle Library, a 1906 piece of pastry erected thanks to the $220,000 donated by Carnegie, a building that was immediately too small, as was the International Style abomination that replaced it in 1960, now razed and forgotten beneath the weight of the flamboyant glass racecourse stand by Rem Koolhass for $165.5 million, which opened recently. Visitors are flooding in, but are they reading? During this same time, in the new reading room of the New York Public Library, anonymous readers—in accordance with the royal principle of this great establishment—have been coming with their laptops to take advantage of the free high-speed connection and are not asking for any works in paper. One might well imagine they are busy consulting them directly from all over, including those that can be found beneath their feet in the stacks of the NYPL. If they still consulted physical books.

William J. Mitchell, professor of architecture as well as director of the Media Lab at MIT: "The facade of the library is not to be constructed of stone and located on a street in Bloomsbury, but of pixels on screens scattered throughout the world. [And, as far as I'm concerned], there is nothing left to put a grand facade on."[1]

This observation was uttered at the beginning of the 1990s. But for politicians to grasp a truth, there need to be at least one hundred deaths.

Appendices

APPENDIX ONE

Libraries and Digitization

This is the day the world changes.

JOHN WILKIN, LIBRARIAN AT THE UNIVERSITY OF MICHIGAN

Since December 14, 2004, the book world has been buzzing with alarm and excitement in the wake of the announcement by Google Incorporated, "part of whose mission to organize the world's information"—a message that has become "whose mission is to organize . . ." as of the day after tomorrow, because the will to power resembles boiling milk—of the agreement recently reached by this company with five major libraries (four American and one British) to digitize their books and put their content online in a form that is not only readable but also searchable and usable by the researcher. For some time the worm of Google has been silently burrowing into the fruit of all the great librarial sites, the BnF included, by disbursing annual checks of significant size; the Mountain View search engine equips in particular the ALA [American Library Association] octopus quite efficiently. Typing "Patriot Act" on February 16, 2005, the searcher found 1,509 articles. Furthermore, a new version of the "engine" indexes and ranks the articles of scientific and university publications under the name of Google Scholar, already nicknamed "Schoogle" by its users. The only ones finding any cause for surprise in this story would be ostriches.

This is how tens of millions of books in the public domain are going to be combined in a common effort, if not a common purpose, by Google Print, which offers extracts and bibliographical information on books still under copyright—"a little like browsing through a new book in a bookstore," the press attachés exulted, assuring the publishers caught by surprise that they had nothing to lose from this service, provided free of charge for the moment, which would allegedly increase their audience and attract new buyers toward their sites and to bookstores. But it is easy to imagine that Google Print would not stick for long to the criteria introduced at its launch and would quickly get to the real gist of the matter: to offer reading for a fee and the opportunity to print copies of the books one wants. (Isn't the very name of the service more or less an indirect admission of this?)

Is the huge immaterial library truly a brilliant idea? Yes, insofar as it involves mountains of advertising dollars falling into the Californians' already rich coffers, thanks to capital—the world's literature and science—that they are acquiring for nothing. Or what is almost nothing for them. While digitizing billions of pages requires a process whose pioneers (Project Gutenberg, Gallica, Library of Congress, and the British Library) are fully aware of its molasses-like pace and its exorbitant cost, Google has been promoted to unfettered magnate since its introduction on the stock exchange, an expense of $100 or $500 million can hardly put a strain on its treasury. We still do not know—as Google itself may not—all the technical details of this project, but it seems that it is beginning with the perfection and large-scale production of rapid automatic scanners, of which the Kirtas Robot is an example (1,200 pages an hour, $70,000 per machine), as well as the training of specialized personnel (although the business managers swear that the operation is as simple as changing the ink cartridge in a copier), with the works being digitized to be placed permanently in the large archives that signed this agreement. In a word, the whole affair must be dealt with briskly.

This benefit for humanity abruptly opens the door to all kinds of disarming consequences and unknowns. Here is an initial collection.

- The intellectual advantage is immediately obvious. The great dream of the universal and encyclopedic library that it pleases us to impart to the Ptolemies is in the process of being realized right there on the very table in front of you. It truly facilitates the labor involved in university and literary research, which generally consists of not reading long portions of text or entire books.
- The handling of old books can endanger them; putting them through a high-speed automatic process seems even more perilous. Uncertainty turns into anxiety when one considers the pages that have turned brittle. This would include a large portion of the books printed between 1850 and 1950.
- Unless we imagine that the invaders from Silicon Valley have been smoking pot, their intentions are quite far from free distribution of reading material. No one, in fact, is yet capable of devouring a book on the screen. The page is not displayed in its entirety and demands ceaseless gymnastics from the reader even on a large monitor, which includes the gesture for "turning" the page, not to mention the times spent going backward, browsing the text, consulting the front and back matter, and so on. Even when these contortions with the mouse that are auspicious for triggering pop-ups (untimely advertisements that smack you right in the eye) are not taken advantage of, the eventual appearance of ads will be a prerequisite for access to this service at the same time the service will be authorized only for registered users—in other words, users who have been listed and paid their fees, and thus are likely to be "informed" of multiple propositions selected according to their profile. The display of the desired Web pages will then be accompanied by various immediately useful links that

are particularly attractive to the quickly overwhelmed apprentice reader, such as the closest library where the book in paper can finally be read—although a number of librarians are secretly and desperately hoping to take advantage of its digitization to retire it from circulation—or as a link to eBay, Amazon, or Abebooks in order to effortlessly acquire it, either new or used. This process is exactly identical to that of Google Print.

- A secondary question: Since offering content seems so vital, are Yahoo, MSN, and others going to launch themselves on the same course? Make deals with other archives? Perhaps they will offer the same collections and indulge themselves in contests like this: "Discover Proust and win a Twingo!"*

Confronted by so many questions, Harvard Library prudently took pains to specify that the operation would first focus its efforts on forty thousand of its fifteen million titles as a test, and that every precaution would be taken for their safety. A communiqué couched in terms as evasive as possible affirming absolute confidence in the usefulness of this project could not hide the extent the game master is now an uneducated and uncontrollable authority: "Unlike traditional digitizing projects—such as those carried out under the University's Library Digital Initiative or the Open Collections Program—which are based on careful book-by-book selections of the very best or most appropriate materials on a specific topic, Google's approach is simply to digitize as many books as possible." Oxford has more drive, predicting that a million books from the Bodleian will be available online in three years (with the exception of rare and valuable works, for the moment the exclusive property of its own digitization department). Striking a similar note—in fact, singing the same tune—is New Yorker Paul LeClerc (NYPL): "It's a significant opportunity to bring our material to the rest of the world. It could solve an old problem:

*[Twingo is a "fun" sports car designed by Renault. —*Trans.*]

If people can't get to us, how can we get to them?" (Did this cause Lagardère to prick up his ears?) And John Wilkin (Michigan) goes him one better: "This is the day the world changes. It will be disruptive, because some people will worry that this is the beginning of the end of libraries. But this is something we have to do to revitalize the profession and make it more meaningful."

The choice of terms is often more long-winded than the spokesperson would like. Google feels it is invested with a worldwide "mission" and accepts the shouldering of a colossal "wholesale" labor, words that smack of a desire for vaguely Christian-like hegemony that shares a mercantile mind-set at the same time. And, by means of a press release, this is what we see coming—in the same way the clown announces the circus—the obligatory "No library books were harmed in the making of these digital copies." We do not yet know whether this smacks of planetary neo-silliness or well-concealed humor.

During the earliest days following this announcement, the bibliophile Web forged the label "omnigoogolization" to describe the event. Among other mocking observers, a charming family of Indian scholars drew up a list of the winners and losers on its blog. On the left side are the beneficiaries: researchers as well as the public of nonprofessional readers, the holders of Google shares, the countries of the Third World, and lawyers specializing in author copyrights. The losers are, of course, library employees and publishers, the hefty Adobe Reader, which finds it now has an identical twin, and now surpassed precursors like Project Gutenberg (which had already broken down in 2003), Questia, ebrary, Million Book Project, and so on.

With respect to the Third World, this in no way involves the great non-English-language archives like those of France or the Hispanic countries, the Arab world, or the Chinese world. The author of these lines sent a short request to the press secretary at Google Inc. headquarters for information on this subject on December 15, 2004, and again the following January 19, a request that remains unanswered to this day. As they are quite considerable in size, these collections are inevitably—

in the profit-making logic at the base of the operation—desirable. But unless secret negotiations were aborted, they are not desired, at least during the first phase of the investment. The wealth of foreign books in the large Anglo-Saxon research libraries is certainly noteworthy, and placement online concerns them, too. But in any event, the strategy being executed appears to be marginalizing only non-Anglo-Saxon cultural spheres and foreshadows the potential transformation of certain vectors of thought, such as French, into regional languages.

Because the dinosaur lying helplessly on its back with its feet in the air is Tolbiac [home of the new National Library of France], with the help of other matters of concern, six weeks went by before the representative of the BnF, the primary victim, publicly voiced his dismay (*Le Monde,* January 23 and 24, 2005), without, however, taking stock of the announced tsunami. Google is in no way "challenging" Europe, a negligible entity in its eyes. Google is laying hands on the world's knowledge. There are therefore good grounds for fearing that nothing will shock our friends and neighbors, particularly Germany, a country in which *Harry Potter,* before the German translation was published, became the number one bestseller there in its original language: English. So who is left as a culturally respectable ally for France in a possible European endeavor? Spain? But it is also foreseeable that the second series of libraries lusted after by marketing will be Hispanic. France will then find itself all alone, left high and dry on a kind of galley stranded in the sand.

Our country—this is its nobility—has an undeniable weakness and know-how for gasworks. Gallica, in particular, is a colossal investment in most impractical chic, and one day will be admired for having displayed on the screens of millions the expensive pages of public domain text in a policelike and Flaubertian pettiness called "image mode," which prevents you from working.

This "globalization" of the library was foreseen and announced at least modestly in *Books on Fire,* the sole "fill in the blank" being the identity of the operator: we expected Bill Gates; we got Larry Page.*

*[Cofounder of Google. —*Ed.*]

So what can be done when we are confronted by such effective imperialism? Gathering with our Brussels companions to launch a solemn appeal and to consider common actions to take appears overly ambitious, comes too late, and will take years; as for pleading for a cultural exception for Estonians . . .

This new resource scoffs at frontiers and sovereign nations. But this is what true knowledge has always done. For this reason alone it is desirable that the contents of the BnF—in which the majority no longer belongs to anyone, need I remind you—be made, at whatever cost and immediately, part of this gigantic digitization wave, which, despite all the worries that exist presently and those yet to come, will place before the eyes of the world, in just a short time, all the knowledge of the world.

Article published in Le Monde, *February 8, 2005, under the title "La BnF chez Google? Chiche" [The BnF in Google? I Dare You].*

APPENDIX TWO

Is the Library at a Turning Point?

Library with a capital *L* is the archetype of which I am speaking. The general model of a book collection is intended to be encyclopedic and universal, made available to everyone at no charge, with no other obligation than to respect it.

In this sense, its embryonic form appeared at the Museion of the Ptolemies around 300 BCE. It was there that the library acquired the name by which we know it in France [*bibliothèque*], a word declined from *bublos,* the papyrus stem. Alexandria carved out a place in our imaginations with the flames from which it was always reborn. We remember the fire caused by Caesar, but it was harmless. On the other hand, we have eight dates of pillaging, assault, and destruction just between AD 213 and AD 415 alone.

213 Caracalla condemns all Alexandrians to death and eliminates all subsidies for the museum

272 Aurelian invades the city occupied by Zenobia; the royal quarter is ravaged; the scholars frequenting the Serapeum flee the country

296 Diocletian puts Alexandria to fire and blood after an eight-month siege

297 or 298	He has all the ancient books of the Egyptians destroyed that are capable of helping them manufacture gold
303	Galerius, the instigator of Diocletian, has the writings of the Christian prophets burned
362	The daughter library is seized by Constantinople, perhaps burned by Jovian (Botti)
391 (389)	Theophilus razes the Serapeum and, if it still survives, the daughter library
412	Archbishop Cyril launches a murderous campaign against those holding ideas competing with Christianity, mainly the Jews and the final representatives of Hellenism. Hypathia is murdered in 415.

Indeed, since the very beginning of its history, the Library has drawn down every kind of misfortune upon itself.

The first source of its woes: the intrinsic blend of pride and fragility.

The Library is arrogant: it displays its superiority by proclaiming it contains something more precious than gold, knowledge. The individual passing before its walls is crushed by it, the individual living in the neighboring country is jealous of it. Both must then either enter and acquire the learning it possesses peacefully or destroy it physically.

The Library is weak: unlike large collections of paintings, it is not easy to move out of harm's way when threatened by fire or rising water, and the individual value of books is too modest. They must all be moved or none; a handful of incunabula does not a Library make.

Thus it must remain in place when danger threatens, and its servitors rarely put their own lives in peril to defend it.

Outside danger has several masks at its disposal.

The most anodyne of these masks is that of accident: fire, flood, and volcanic eruption are not lacking in the chain of events, from the fire that ravaged Anna Amalia on September 2, 2004, or the Capitole with the libraries of Octavian's portico in AD 80, to the modest Sri Lankan collections carried off by the waves during the recent tsunami.

Immobile libraries have suffered from accidents like these thousands of times through history. American insurers have listed three hundred fifty-nine disasters of this nature in their territory between 1911 and 1961.

Grotesque is the mask that preaches hate against book collections, one of a bestiality consented to and proclaimed as such, from the Mongols investing Baghdad to Goebbels burning books, or else the Peasant Rebellion of 1524 in Germany, or even the numerous attacks on libraries in India on a regular basis even now, particularly research or university libraries. We have Poona in January 2004 and then the library of Imphal in the Manipur region on April 13, 2005.

Finally, there is another mask that expresses, with a dreadful serenity, the subtle hate of ideas deemed dangerous, as is the case with the Inquisition over the centuries.

As a general rule, monotheism displays a noteworthy record for bibliocide throughout history. If there was a contest held on the number of books burned and libraries destroyed, Catholicism should be the first to mount the podium, followed by Islam. But there is also the new religion—Stalinist-brand Communism—created from whole cloth in Moscow around the middle of the last century, the final and colorful remnants of which can still be observed in places like Havana.

Just like the Nazis in Czechoslovakia and Poland, Stalin decided to eradicate local memory in the Crimea, Estonia, and other regions under his thumb. This is also what the Spanish did in Mexico in 1529 and in the Yucatán in 1561. In all these cases, there has been perceived a pressing need to carry the destruction beyond just human beings. In order (at the very least) to get rid of any material and moral claim once and for all, it is necessary to erase all cultural traces.

Biblioclasty has therefore managed several times in history to attain that unsurpassable peak of achievement consisting of the thoroughly planned decision to bring about the disappearance of not just one but all the libraries of a community.

People born after the Second World War sincerely believed that this was something they would never see again.

However, this is precisely what happened on August 25, 1992, to the National Library in Sarajevo, then to two hundred libraries that the federal Serbian troops systematically destroyed or allowed to deteriorate for purposes of ethnic cleansing.

It is possible that we are again witnessing this kind of collapse close to home. The incompetence of the international community during such events and its absence afterward, when repairs are required, do not give us good grounds for optimism.

Despite all these snubs, the model for the large text archive lingered on and multiplied over three millennia.

After the Enlightenment, society put effort into respecting and maintaining this common property. Therefore we now have before our eyes—and we happily use them—an admirable number of rich and well-managed institutions, some of which are colossal in size, although they are not National Libraries, from New York to Shanghai, down to the myriad of specialized collections in every country of the world.

Yes, constant conflagrations, wholesale pillaging, and the Inquisition now seem to belong totally to the past.

But does this mean that the Library is still left in peace?

Are we not rather in the midst of inventing troubles "of an entirely new type" that will be even harder to overcome?

In the first place, since 2003, a considerable torrent of local libraries are shutting down or are being forced to resort to major reductions in hours and, at the minimum, operating without a fully trained staff. This phenomenon began in the United States, where public reading is dependent upon foundations and directly suffers from any reduction of stock earnings on their capital. Hundreds of libraries will not be open after 2006.

But the same phenomenon is occurring, it appears, in South Africa as well as in Germany. In France, three occurrences of this nature have recently been detected—small in scope, certainly, but all within the space of a single month (the headquarters of INSEE [Institut National de la Statistique et des Études Économiques], the BCE of Renault Le Mans, and the Library of the Employees of the City of Paris). In Brussels, the Ministry of Health closed the Vesalius Library and scattered, some even say destroyed, its contents. There are then the cases of the all-powerful New York Public Library, which decided to sell its paintings to make up for revenue shortfalls; and that of the sacrosanct Lambeth Palace Library of London, which has been called upon the accountant's carpet for the first time since 1610 and risks eviction.

A measure adopted by communities in France for continuing to finance public reading has been to modernize it. The recipe is quite simple: build a small edifice of smoked glass, place inside the most heavily borrowed books from the old Municipal Library and bestsellers, add to that graphic novels, CDs, and DVDs, plus game consoles and computers connected to the Internet, then rebaptize this savory delicacy a "*mediathèque.*" Alas, the observations beginning to come in on these new establishments are not very reassuring. By virtue of their resemblance to supermarkets, they inspire in people the attitude of a casual consumer to whom everything is owed, and who—if you remove *Harry Potter* and *The Da Vinci Code* from the shelves—read practically none of the books housed there.

This trend of disrespect on the part of the "customer" (that's his or her new title) is also evident, in a way, in the behavior of the sponsor of the *mediathèque.* On April 28, 2005, for example, the mayor of Marseilles requisitioned the brand-new Alcazar, which cost sixty-one million euros, most of it paid by the state, to host a housing convention. Readers and librarians were cavalierly urged to stay home.

If the public authorities do not have any consideration for the Library, they will unblinkingly make the decision to close it once it has become a target for increased vandalism. This is the case for two small establishments in Great Britain, one in Warwickshire and the other in the West Midlands.

Before such distressing and repeated cases, we may ask if keeping the Library alive, continuing to add paper books to the temple of reading, whose upkeep is more and more expensive, is not actually attempting to swim against the current of history. Perhaps making all books available in virtual form to users even in their own homes will be, most often, a solution that comes just at the right time.

If it had been built, the Mundaneum would have been erected between Grand-Saconneux and Pregny—where the League of Nations, then the United Nations, set up shop—to provide, in the words of its architect, Le Corbusier, "the most majestic spectacles on the four cardinal points." And "so that the total image and meaning of the World can be seen and understood on one point of the Globe," concluded Paul Otlet, initiator of the Transnational Library that would never come into being.

This was just like the World Brain imagined by H. G. Wells, which also failed to be physically realized. Based in Barcelona, this planetary encyclopedia, engendered as a permanent entity by universities and research centers, would have employed seventeen million employees located in every country.

These Utopias of 1928 and 1939 were intended to endow human beings with an increase of individualistic awareness in order to save the world from ruin.

Too physical to succeed, will these Utopias become a reality with the scheduled digitization of millions of books announced by Google on December 14, 2004?

Even if they are invoking a "mission" today—although we have no idea what deity invested them with it—the owners of Google are the opposite of Utopians: earning even more money when you are already a billionaire is a purpose that at least offers the advantage of clarity.

The affair got its start in Stanford, where one of Google's founders was a student. His future business partner, from the University of Michigan, would join him there. Their major concern was, and still is, to rediscover information wherever there was too much of it for any classification system to be effective.

It has now been more than two years* that Google, with the technical support of a company based in Ecublens, Switzerland, has been hard at work digitizing Stanford's books. Nor is there anything surprising in the fact that the second library to sign a contract was the University of Michigan.

To refresh our memories, the others are Harvard, the New York Public Library, and the Bodleian Library of Oxford, each representing different degrees of purpose and commitment:

- Stanford: all eight million of the books it owns
- Michigan: the same, seven million
- Harvard: forty thousand books (of fifteen million)
- New York: no announced details as of yet
- Oxford: one million (but not the incunabula and manuscripts, which it is digitizing itself, and nothing after 1901)

As we can see, the approaches are diverse and the quantities vary; the nature of the books is heteroclite. No one can speak here about a campaign plan, unless he or she wants to wage war.

All other information is coming, for the moment, only from journalists' rehashes or, more exactly, bloggers' daydreams. People are peddling the figure of fifteen million titles because this is the result of adding Stanford to Michigan; people speak of a ten-dollars-per-copy cost because this is the figure the *New York Times* printed on December 16, 2004, based on an anonymous prognostication, which, when one knows what libraries have been spending until now—it is more like one hundred dollars and even one hundred euros—would represent rather a goal to be achieved, without which the operation would not be entirely successful. In other words, attempting to mobilize Europe around a project costing only $150 million smacks of total unawareness or a game of three-card monte.

*[As of 2009, it is five years. —*Trans.*]

It is no time to play word games when speaking about digitized books intended for all readers, including students, faculty members, researchers, and so forth. These books must be searchable in English, indexable, and even "interoperable." Using jargon from the computer science field, the books need to be text mode, not image mode; XML, not PDF. To tell the naked truth, the sixty-six thousand books of Gallica are all to be OCR scanned in order to enter the folds of history and be taken into consideration. Otherwise, they will not exist, and the BnF really does not have anything to offer except for the twelve hundred titles amiably entrusted to it by the CNRS.

But today, on the technical and financial plane, we find current events in a tunnel. Europe signs letters of intent that will be of no consequence, whereas Google has ceased all communication concerning content. Protesting his good faith, Google's spokesman turned on the charm and refused to directly answer any concrete questions, as was the case at the Twelfth National Conference of the ACRL (Association of College and Research Libraries) in Minneapolis on April 10, 2005, or in Troyes, France, on June 3 that same year. Let's note in passing that Paris is more prepared to organize a huge small-business affair in 2012 than to organize a constructive confrontation of the last ramparts of intelligence with the thing that threatens to crumble them into dust forever.

But let's imagine that things have transpired as reported by all the media: to the two or three thousand now available today on the World Wide Web, somewhere between ten and fifteen million titles will be added in five years.

So what are the changes then?

In the first place, Google (or someone else, because its competitors are not just stewing in their corners) will be earning mountains of money in 2010: book pages will be unscrolling across the screens with advertising banners for its own services, for the library where the book in question can be found, and for the bookstore; and then, contrary to what was announced in 2004, for all the advertisers who can afford to pay enough for this service.

The second point, and the most important, is the constructive accessibility of books in places without a library and where there is not a chance, for decades to come, for those places to create their own decently equipped libraries. It is fine to provide for remote places like Lomé in Togo, but we need to take care of our own rural areas as well. Doing so will derive another benefit, which is the great facility to offer professional writers and researchers whose vocations depend on reading and writing. They will no longer have to leave home to perform their work, or will have to travel only to the nearest library when they cannot find what they need online. Among other factors, this last motivation will guarantee the future survival of old libraries.

The third consequence is serious, but one that also remains in the category of ungrounded hypotheses and predictions. Will cataloging books online be possible, or even conceivable? Will we continue to be satisfied with the customary algorithm that consists of listing metadata in the order of the popularity they've managed to garner up to any given moment? In this reckoning, we are not even talking about a "library" anymore, but rather an astronomical dictionary of quotations. Can I type out part of a sentence and harvest what I need to write my dissertation? But only those who totally turn their back on the Internet can imagine it constitutes a modern Quid* in Limbo, in which a kind of ill-intentioned will or enemy tongue would exercise some sort of supremacy. In reality, it does not evolve like an infinitely expanding galaxy in which nothing is given a ranking or lost. Everything lingers there in the wings, and the sole manner of profiting from it is to pose an intelligent and detailed question—a talent quickly learned.

The fourth and final observation: the snowball. "Because we will be able to do it, we will continue doing it!" The necessary machines will be perfected, digitization will become cheap, and all institutions will benefit from it, therefore bringing into the game all the non-English-language

*[Name of a French online encyclopedia. —*Trans.*]

collections that might still be missing from the great online library at that time.

As the *content* of books fallen into the public domain does not belong to anyone, any claim of sovereignty concerning them seems completely out of place. First of all, while it may be a majority of English-language books that prove to be available on the Web, the sole responsibility lies with those who have deplored this state of affairs most loudly, and who held the authority in Paris to launch the digitization of millions of French books in 1988, then again in 1994, but who preferred to construct a monument undoubtedly worthy of condemnation in both meanings of the word. Furthermore, it is inappropriate for these authorities to attribute their own defects to Google, which is less America-centric than dollar-centric. There are, in fact, 324,000 French books—true classics, not minor titles—in the Stanford collection that are joyfully departing for the Great Digitization.

Fortunately, we have had a European Library since March 17, 2005, in other words, a portal that will allow for rapid discovery of the forty-three principal archives of the Continent and whatever they wish to contribute to what almost amounts to an *auberge espagnole.** But it is not this excellent initiative, no more than it will be a motion signed by nineteen countries, the most indigent of which were in first, for "an energetic research program in the technical domains that will be used to achieve this plan," that will give us the billion euros necessary for the realization of the belated dreams of our old leaders: a search engine and huge digitized European library in which, if I understand correctly, Shakespeare, Swift, Byron, Oscar Wilde, Lewis Carroll, and the Brontë sisters would not be admitted in their native tongue.

Now that the jeers about blind French arrogance are proliferating

*[This term translates literally as "Spanish inn," but *auberge espagnole* means a place where guests have to bring their own meal, resulting in the idiomatic expression of a cultural "melting pot." —*Trans.*]

on the blogs of academics and librarians everywhere, I do not know if it will be possible, as was undoubtedly the case before January 22, 2005, to take advantage of the digitization service that had been offered. Then the expression "Google challenges Europe," beyond the fact that it smacks of the same kind of offensive mendacious advertising publishers use to sensationalize their books, only tosses oil on a fire that no reasonable person needs to cook up the future.

As Paris is now perhaps damaged goods, shouldn't another European capital endeavor to represent our desire for knowledge and, by adopting a more pragmatic, universal, and neutral attitude, invite Google's engineers to install its machines and show us what they are capable of doing? In granting the latter the benefit of the doubt, all of us, as the world's readers, have everything to gain.

But in parallel, the final avatar of the curse of the large Library is on the march: it is the commercialization of all branches of learning connected to an inordinate juridism; it is locking up intellectual property under the hypocritical pretext of transferring it from paper to screen, a case file in which it seems that France is getting ready again to distinguish itself by applying the harshest restrictions in the whole of Europe to reading for free, while ignoring the grumbling that emanates from Victor Hugo's tomb. Don't we know any better, he might ask, than to point the finger at a businessman from the New World who is only trying to do his job, whereas we are turning down the honor of taking steps to ward off new dangers, dangers that are much more vicious and potent than those that threatened us in the past?

Yes, even if we take numberless old horror stories into consideration, the Library now finds itself at a historical turning point, and the most difficult stage of its journey perhaps still lies ahead.

This speech was given in Geneva on April 29, 2005, by invitation of the Swiss National Library.

Notes

Many of the subjects in this book could not be examined without documentation that is available only on computer networks, hence the considerable number of links that I've provided here as references. To facilitate any use one might want to make of them, the reader is invited to go to the author's website and click on them directly from this page: www.polastron.com/notesLGN.html.*

INTRODUCTION: FINDING INFORMATION OR FUNDING A NATIONAL LIBRARY?

1. Investigation headed by Senators Philippe Nachbar and Philippe Richert, "La Bibliothèque nationale de France: un chantier inachevé" [The National Library of France: An Unfinished Worksite], Information report no. 451 from the commission for cultural affairs of the Sénat, June 2000, whose publication was greeted with remarkable discretion by the media, who have a tendency to cite only books on the National Library written by members of its former management. Two books provided a much less moderate response: Jean-Marc Mandosio, *Après l'effondrement* [After the Collapse] (Paris: Encyclopédie des Nuisances, 2000), and Jean-Pierre Guéno, *Petites Chroniques de l'amnésie ordinaire* [Brief Chronicles of Ordinary Amnesia]

[*Despite all best efforts to keep current, given the speed of the Internet's constant mutation it is possible that some of these links will have expired by the time this translation is available in bookstores. —*Trans.*]

(Toulouse: Éditions Milan, 2004). The entire report can be read online at: www.senat.fr/rap/r99-451/r99-451html.

2. www.bnf.fr/pages/infopro.cooperation/po_chartegallica.htm.
3. Dominique Lahary, librarian, voicing his personal position on Biblio.fr on May 1, 2005.
4. *Culture et Recherche,* no. 100, January–March 2004, which can be found online at www.numerique.culture.fr.

CHAPTER 1. BnF VERSUS BNF

1. Jean-Noël Jeanneney, "Quand Google défie l'Europe" [When Google Challenges Europe], *Le Monde,* January 23–24, 2005, and Lucien X. Polastron's reply, "La BnF chez Google? Chiche" [The BnF in Google? I Dare You], *Le Monde,* February 8, 2005. The untimely appeal of the BnF's spokesman called for at least some response. As no one was close to taking on, or even wished to take on, this responsibility, *Le Monde* wanted to publish the pages written for this book after December 16, 2004, accompanied by an independent opinion. In addition to the inclusion of a title it coined itself and giving Jeanneney the first name of Jean-Michel, this newspaper made several excusable but awkward censorings. Appendix 1 therefore includes a copy of this article as it was sent to *Le Monde.*
2. Frédéric Lepage detected its presence on the Internet as follows: "Based on a notice from the national library: collection of the legislation of the Committee of Public Safety with the official correspondence of its representatives and the registry of the Provisional Executive Council. . . . Reproduction: BNF Number of the edition (no place cited): Pergamon Press, cop. 1990. 9 microfiches. The archives of the French Revolution=The French Revolution Research Collection." This is almost all the documents relating to the French Revolution that have been digitized from the microfiches published by Pergamon Press while Paris was hashing out the details for the Bicentennial festivities.
3. *Conversations de Goethe avec Eckermann* [Goethe's Conversations with Eckermann] (Paris: Gallimard, 1988), 206.
4. *Livres Hebdo,* no. 600 (May 6, 2005).

CHAPTER 2. FOR WELLS IS NOT THE PLURAL OF ORWELL

1. Herbert George Wells, *World Brain* (London: Methuen and Company, 1938); Dave Muddiman, "The Universal Library as Modern Utopia: The Information Society of H. G. Wells," *Library History,* vol. 14, no. 2 (November 1998); Warren W. Wagar, *H. G. Wells and the World State* (New Haven, Conn.: Yale University Press, 1961).
2. Antonina Vellentin, *H. G. Wells ou la Conspiration au grand jour* [H. G. Wells, or Open Conspiracy] (Paris: Delamin et Boutelleau, 1952).
3. This expression first appeared in Gottfried Mayer-Kress and Cathleen Barczys, "The Global Brain as an Emergent Structure from the Worldwide Computer Network and Its Implications for Modeling," *The Information Society* 11 (1995), 1–27 (Indiana University, Bloomington).

CHAPTER 3. GENEVA: WORLD CAPITAL

1. Paul Otlet, Mundaneum, no. 128 [by Paul Otlet, author of the text, Le Corbusier, and P. Jeanneret] (Brussels: Union of International Associations Publication, 1928); Paul Otlet, *Traité de documentation: le livre sur le livre: théorie et pratique* [Treatise on Documentation: the Book on the Book, Theory and Practice] (Brussels: Van Keerberghen, 1934) and (Brussels: Éditions Mundaneum-Palais mondial, 1989). The archives of Paul Otlet and the Mundaneum may be consulted at the French Community Archive Center, 76, rue de Nimy, B-7000 Mons, telephone: 065 31 53 43; e-mail: info@mundaneum.be; website: www.mundaneum.be.

CHAPTER 4. QUICK, QUICK

1. Robert McHenry, "The Faith-Based Encyclopedia," *Tech Central Station* (November 15, 2004). www.techcentralstation.com/111504A.html.
2. Jean-Baptiste Soufron, "The Political Importance of the Wikipedia Project: The Only True Encyclopedia of Our Days. Wikipedia: Towards a New Electronic Enlightenment Era?" (November 16, 2004), http://soufron.free.fr/spifron-spip/article.php3?id_article=71; and also "L'encyclopédie libre Wikipedia face aux questions de crédibilité" (September 6, 2004), http://soufron.free.fr/spifron-spip/article.php3?id_article=57.

CHAPTER 5. VOLUTES

1. Cited by Laurence Sanantonios, journalist for *Livres Hebdo,* in a book whose title could also serve as an exorcism: *Tant qu'il y aura des livres* [For So Long as There Are Books] (Paris: Bartillat, 2005), from which also comes the figures for "weeding" mentioned in the sentence that follows.
2. Alain Patez and Pascal Schmitt, "Bibliothèque et lecture en mobilité" [Library and Reading on the Move], *BPF,* vol. 49, no. 6, Paris (2004): 98–104. http::// bbf.enssib.fr/bbf/html/2004_49© 6/2004-6-p98-patez.xml.asp.
3. TextArc is located at www.textarc.org. The pages of manuscripts can be turned with the aid of a mouse on: www.bl.uk/omlimegallery/ttp/digitisation/ html. One can follow Pierre Cubaud's work on http://jasmin.cnam.fr:8081/ as well as http://cnum.cnam.fr/.

CHAPTER 6. A DIGITAL CORONARY

1. Michael Gorman, "Google and God's Mind: The Problem Is, Information Isn't Knowledge," *Los Angeles Times,* December 17, 2004, www.latimes.com/ news/opinion/commentary/la-o e-nugorman17dec17,1,7568022.story?coll-la-news-comment-opinions; Michael Gorman, "Revenge of the Blog People!" Library-Journal.com (February 15, 2005), www.libraryjournal.com/article/ CA502009?display=BackTalkNews&industry=BackTalk&industryid=376 &verticalid=151&; poll of the *"blogbrarians"*: http://dmoz.org/Reference/ Libraries/Library_and_Information_Science/Weblogs/.

CHAPTER 7. WHEN THE BOOK IS TOO HIGHLY CONCENTRATED, THE PURPOSE IT SERVES IS EASILY FORGOTTEN

1. *Esprit,* "Malaise dans l'édition" [Publishing Is Under the Weather], June 2003; André Schiffren, *Le Contrôle de la Parole* [The Monitoring of Speech] (Paris: La Fabrique éditions, 2004), his sequel to *L'Édition sans éditeurs* [Publishing without Publishers], 1999. Yet even more depressing: Janine and Greg Brémond, *L'Édition sous influence* [Publishing Under the Influence], 2nd ed. (Paris: Éditions Liris, 2004).
2. The expression describes one of the four winds on the frescoes of the Torre dei Veniti at the Vatican. Pope Alexander VII hastily had it covered over

by green paint when Christina of Sweden visited in 1655. Verena von der Heyden-Rynsch, *Christine de Suède: La souveraine énigmatique* [Christina of Sweden, the Enigmatic Sovereign] (Paris: Gallimard, 2001), 110.

3. Gary Wolf, "The Great Library of Amazonia," *Wired,* issue 11–12 (December, 2003). www.wired.com/news/business/0,1367,60948,00.html (and since its appearance four new titles have been added, proof that this Boss Tweed–like scandal is inspiring interest).
4. Samuel Blumenfeld, "Le Noé du Yiddish," *Le Monde 2,* September 25, 2005.
5. www.yiddishbookcenter.org/+10055.
6. Remarks by Nigel Newton of Bloomsbury and quotations from the professional journal *The Bookseller,* reported in *Livres Hebdo,* no. 559 (April 29, 2005).

CHAPTER 8. THE PIXEL COMING TO PAPER'S AID

1. M. J. Rose, "Self-Publish Stigma Is Perishing," *Wired News* (July 23, 2003), http://wired-vig.wired.com/news/culture/0,1284,53996,00.html.
2. To see its catalog, order a book, or simply browse, go to www.nap.edu/about.html.
3. *Tant qu'il y aura des tomes,* the files of the *Canard enchainé* (Paris: 2004).

CHAPTER 9. IS THIS ALREADY THE POST-GOOGLE ERA?

1. *Le Point,* no. 1720, September 1, 2005, p. vi. It has become increasingly easier to follow the topicality of this case file, made even more captivating by the obstinate silence of the main party, thanks to reviews in the press; in French: www.lapresseaffaires.com.

CHAPTER 10. BUT WHY THE DEVIL DO WE NEED LIBRARIES?

1. Kirtas Technologies Inc. in Germany: http://kirtass.tech.com
   4DigitalBooks in Switzerland: www.4digitalbooks.com
   Infotechnique in Strasbourg: www.infotechnique.com
   Spigraph Images: www.spigraph.fr/

CHAPTER 11. CONCORDANT AND DISCORDANT CLUES

1. Adam Smith (1723–1790), *Enquête sur la nature de la richesse des nations* [An Inquiry into the Nature and Causes of the Wealth of Nations] (Paris: PUF, 1995).
2. "Electronic Voltaire" is a "product" of ProQuest Information and Learning, the American representative of the British firm Chadwyck-Healey, which describes its mission as one serving the information and teaching fields; www.chadwyck.co.uk/.

CHAPTER 12. THE BIG PICTURE

1. John Markoff and Edward Wyatt, "Google Is Adding Major Libraries to Its Database," *New York Times,* December 14, 2004.
2. There were 130 people at the April 27, 2005, meeting of the SNE. Hervé Hugueny, "Google lance son operation seduction" [Google Launches Its Operation Seduction], *Livres Hebdo,* no. 600 (May 6, 2005). The legal representatives of the SNE, on the other hand, never responded to my written request for comment on this subject.

CHAPTER 13. FIRST TRIALS

1. The AAUP letter can be found at www.aaupnet.org/aboutup/issues/0865_001.pdf. Google's declarations can be read at: http://googleblog.blogspot.com.
2. A young law professor named Tim Wu has given much scrutiny to the many legal questions posed by the Internet, copyright, and bestsellers: www.law.Virginia.edu/lawweb/Faculty.nsf/FHPbI/6697.
3. The Authors Guild Message can be found at www.authorsguild.org/news/charity_handy_talking.htm. David Youngberg commentary: http://lawlegislationandlunacy.blogspot.com/2005/09/when-pen-picks-up-sword.html; and Adam Smith's: http://googleblog.blogspot.com /2005/09/buzz-about-google-print-ans-lawsuit.html.
4. www.publishers.org/press/index.cfm.

CHAPTER 14. BURNING STAKES

1. Olivier Ertzchied teaches at URFIST (Unité régionale de formation à l'information scientifique et technique). "Look in the wrong place and find the right thing: Serendipity and information research" can be found at http://archivesic.ccsd.cnrs.fr/documents'archives0/00/00/06/89/sic_00000689_02/sic00000689.html.

CHAPTER 15. ADVENT EVE

1. "Une nouvelle bibliothèque. Un nouveau système d'information" [A New Library. A new Information System], 62nd IFLA General Conference, August 25–31, 1996.
2. Alain Jacquesson and Alexis Rivier, *Bibliothèques et documents numériques. Concepts, composantes, techniques et enjeux* [Digitized Libraries and Documents: Concepts, Components, Techniques, and Stakes] (Paris: Éditions du Cercle de la Librairie, 1999, 2005).
3. Michel Foucault, "Qu'est-ce qu'un auteur?" [What Is an Author?] (1969), *Dits et Écrits* [Things Said and Written] vol. 1 (Paris: Gallimard, 1994).
4. Bernard Cerquiglini, "Variantes d'auteur et variance de copiste" [Variants of Authors and Variance of Copyists], *La Naissance du texte* [The Birth of the Text], Louis Hays ed. (Paris: José Corti, 1989).

CHAPTER 16. *MUTA SOLITUDO*

1. Arundell Esdaile, *Manual of Bibliography* (London: Allen & Unwin, 1967), 57.
2. Jean-Noël Jeanneney, *Quand Google défie l'Europe* [When Google Challenges Europe] (Paris: Fayard, Éditions Mille et une nuits, 2005).

CHAPTER 18. THE FUTURE AT THE PORTAL

1. *Le Monde,* September 1, 2005.
2. www.minervaeurope.org/events/reding050621.pdf.

CHAPTER 19. TOMORROW'S READERS

1. Pierre Corneille, *Agésilas,* Act III, Scene 1, as found, for example, on Gallica: "Votre intérêt s'y mêle en les prenant pour gendres; / et si par des liens si forts et si tendres / vous pouvez aujourd'hui les attacher à vous, / vous les donnez plus qu'a nous." [Your own self-interest becomes involved by taking them for genders / and if by such strong and tender bonds (links) / you can attach them to you today / you give them more than you give us].
2. André Breton, "Caractères de l'évolution moderne et ce qui en participe," *Les Pas Perdus* (Paris: Éditions Gallimard, 1924, 1990). Available in English translation as "Characteristics of the Modern Evolution and What It Consists Of" in *The Lost Steps,* trans. Mark Polizotti (Lincoln: University of Nebraska Press, 1996).

CHAPTER 20. LAST BOOKS! LAST BOOKS! CLOSING TIME!

1. http://librarydust.typepad.com/library_dust/2004/11/Salinas_library.html.
2. www.libraryjournal.com/article/CA506986?display=breakingNews.
3. www.lambethpalacelibrary.org.
4. www.rwevans.co.uk/libri/.
5. See his quadrilingual site: www.health.fgov.be/biblio/.
6. "Questions et Réponses," Senat, May 24, 2005, www.senate.be/wwwcgi/get_pdf?50334058.
7. Christine Garin, "Inquiétudes pour la bibliothèque du personnel de la Ville de Paris" ["Worries about the Library of the Personnel of the City of Paris"] *Le Monde,* February 11, 2005.
8. Marie-Dominique Mouton, "Donner à penser," ["Food for Thought"] *Vacarme* [Loud noise, din] no. 32 (Summer 2005).

CHAPTER 22. LIBRARY ARISE!

1. Benjamin Franklin, *The Autobiography of Benjamin Franklin* (New York: Touchstone, 2003), 56.
2. Victor Hugo, *Discours d'ouverture du Congrès littéraire international. Le Domaine publique payan* [Opening Speech at the International Literary Congress: Public Domain for a Fee] (Paris: C. Levy, 1878).

3. Jacques Drillon, "Culture: Tout se paie, tout se plaide. La privatisation du monde" [Culture: Everything Has a Price, Everything Is in Litigation. The Privatization of the World], *Le Nouvel Observateur,* no. 2117, June 5, 2005.
4. www.wipo.int/about-wipo WIPO, World Intellectual Property Organization; WCO, World Commerce Organization; the law prohibiting the personal copy is the DMCA, Digital Millenium Copyright Act.
5. www.bmlisieux.com/maj.htm.

CHAPTER 24. SHROUDS

1. Laurence Santantonios, *Tant qu'il y aura des livres* [As Long as There Are Books] (Paris: Bartillat, 2004).
2. Jérôme Lindon, "De l'édition sans editeurs" [On Publishing without Publishers], *Le Monde,* June 9, 1998.
3. The ADBS, or Association of Professional Information and Documentation Workers, offers a very clear file on this thorny legal question: www.adbs.fr/site/publications/droit_info/droit_prêt.php#directive.
4. Baptiste-Marrey, *Prêter* (*in livre*) *n'est pas voler* (*son auteur*) [Lending (a Book) Is Not Stealing (from Its Author)] (Paris: Éditions Mille et une nuits, 2000).

CHAPTER 25. PURSE STRINGS AND POLICE CORDONS

1. Christian Vandendorpe, "Pour une bibliothèque virtuelle universelle" [For a universal virtual library]. *Le Débat,* no. 117 (November–December 2001). Available online at www.diplomatie.gouv.fr/culture/livre_et_ecrit/revue/le_debat/117/internet/art_01.html.
2. Daniel Zimmerman, *Alexandre Dumas le Grand* (Paris: Phébus, 2002).

CHAPTER 26. SMOCKS

1. "La notion de 'zone grise'" ["The Notion of 'Gray Zone'"], *ADI, Actualités du droit de l'information,* no. 59, June 2005, analyzes the fuzziness and the impossibility of applying this idea. Association des professionals de l'infor-

mation et de la documentation (ADBS), 25, rue Claude-Tillier, Paris, 12th arrondisement. E-mail: adbs@adbs.fr.

CHAPTER 27. AGAINST THE GRAIN

1. "The facade of the library is not to be constructed of stone and located on a street in Bloomsbury, but of pixels in screens scattered throughout the world. [And, as far as I'm concerned], there is nothing left to put a grand facade on." William J. Mitchell. *City of Bits* (Cambridge, Mass.: MIT Press, 1995), 56. "And if I had to revise this book ten years later, I would say that the facade is now the home page of Google," he added in a personal letter to the author on September 2, 2005.

Bibliography

Baptiste-Marrey. *Prêter (un livre) n'est pas voler (son auteur).* Paris: Éditions Mille et une nuits, 2000.

Blondeau, Olivier, and Florent Latrice. *Libres enfants du savoir numérique, une anthologie du "libre."* Paris: Éditions de l'Éclat, 2000. This book can be consulted for free at www.freescape.eu.org/éclat/index.html.

Breton, André. "Caractères de l'évolution moderne et ce qui en participe," in *Les Pas Perdus.* Paris: Éditions Gallimard, 1924, 1990. Available in English translation as "Characteristics of the Modern Evolution and What It Consists Of," in *The Lost Steps.* Translated by Mark Polizotti. Lincoln: University of Nebraska Press, 1996.

Brown, John Seely, and Paul Duguid. *The Social Life of Information.* Boston: Harvard Business School Press, 2002.

Castells, Manuel. *Fin de millénaire.* Paris: Fayard, 1999.

Cerquiglini, Bernard. "Variantes d'auteur et variance de copiste," *La Naissance du Texte.* Edited by Louis Hay. Paris: José Corti, 1989.

Chemla, Laurent. *Confessions d'un voleur: Internet, la liberté confisquée.* Paris: Denoël, 2002. This book can be consulted for free at www.freescape.eu.org/biblio/IMG/pdf/confessions.pdf.

Esdaile, Arundell. *Manual of Bibliography.* London: Allen & Unwin, 1967.

Fadiman, Anne. *Ex-libris, Confessions d'une lectrice ordinaire.* Paris: Éditions Mille et une nuits, 2004.

Foucault, Michel. "Qu'est-ce qu'un auteur?" (1969) *Dits et Écrits,* vol. 1. Paris: Gallimard, 1994.

Franklin, Benjamin. *The Autobiography of Benjamin Franklin.* New York: Touchstone, 2003.

Gattégno, Jean. *La Bibliothèque de France à mi-parcours. De la TGB à la BN bis?* Paris: Éditions du Cercle de la Librairie, 1992.

Guéno, Jean-Pierre. *Petites Chroniques de l'amnésie ordinaire.* Toulouse: Éditions Milan, 2004.

Jacquesson, Alain, and Alexis Rivier. *Bibliothèques et documents numériques: Concepts, composantes, techniques et enjeux.* Paris: Éditions du Cercle de la Librairie, 1999, 2005.

Jeanneney, Jean-Noël. *Quand Google défie l'Europe.* Paris: Fayard, Éditions Mille et une nuits, 2005.

Latrive, Florent. *Du bon usage de la piraterie.* Preface by Lawrence Lessig. Paris: Exils, 2004. This book can be consulted for free at www.freescape.eu.org/piraterie.

Mandosio, Jean-Marc. *Après l'effondrement.* Paris: Encyclopédie des nuisance, 2000.

Mitchell, William J. *The Logic of Architecture: Design, Computation, and Cognition.* Cambridge, Mass.: MIT Press, 1990.

———. *City of Bits: Space, Place, and the Infobahn.* Cambridge, Mass.: MIT Press, 1995. This book can be consulted for free at http://homepage.mac.com/bogronlund/3_314Upliad_05_03/Mitchell_City%20of%20bits.html.

Muddiman, Dave. "The Universal Library as Modern Utopia: The Information Society of H. G. Wells," *Library History,* vol. 14, no. 2 (November 1998).

Otlet, Paul. *Mundaneum* [by Paul Otlet, author of the text, Le Corbusier, and P. Jeanneret]. Brussels: Union des associations internationales, no. 28 (1928).

———. *Traité de documentation: le livre sur le livre: théorie et pratique.* Brussels: Éditions Mundaneum-Palais mondial, 1934, 1989.

Papy, Fabrice, ed. *Les Bibliothèques numériques.* Paris: Éditions Hermès Science, 2005.

Perec, Georges. *Penser/Classer.* Paris: Éditions du Seuil, La Librairie du XXIe siècle, 2003.

Polastron, Lucien X. *Le Papier, 2000 ans d'histoire et de savoir-faire.* Paris: Imprimerie nationale, 1999 (Actes Sud).

———. *Livres en feu: Histoire de la destruction sans fin des bibliothèques.* Paris: Éditions Denoël, 2004. Available in English translation as *Books on Fire: History of the Endless Destruction of Libraries.* Translated by Jon E. Graham. Rochester, Vt.: Inner Traditions, 2007.

Santantonios, Laurence. *Tant qu'il y aura des livres.* Paris: Bartillat, 2005.

Stasse, François. *La Véritable Histoire de la grande bibliothèque.* Paris: Éditions du Seuil, 2002.

Vallentin, Antonina. *H. G. Wells, ou la Conspiration au grand jour.* Paris: Delamain et Boutelleau, 1952.

Wagar, W. Warren. *H. G. Wells and the World State.* New Haven, Conn.: Yale University Press, 1961.

Wells, Herbert George. *World Brain.* London: Methuen and Company, 1938.

Zimmermann, Daniel. *Alexandre Dumas le Grand.* Paris: Phébus, 2002.

Index